Cladistics:
Perspectives on the Reconstruction of Evolutionary History

CLADISTICS:

Perspectives on the Reconstruction of Evolutionary History

PAPERS PRESENTED AT A WORKSHOP ON THE THEORY AND
APPLICATION OF CLADISTIC METHODOLOGY,
MARCH·22–28, 1981, UNIVERSITY OF CALIFORNIA, BERKELEY

Edited by
Thomas Duncan
and
Tod F. Stuessy

New York   Columbia University Press   *1984*

Library of Congress Cataloging in Publication Data

Workshop on the Theory and Application of Cladistic
Methodology (1981 : University of California,
Berkeley)
Cladistics : perspectives on the reconstruction of evolutionary history.

Includes bibliographies and indexes.
1. Cladistic analysis—Congresses. 2. Phylogeny—
Congresses. I. Duncan, Thomas, 1948–
II. Stuessy, Tod F. III. Title.
QH83.W64 1981 575 83-26178
ISBN 0-231-05430-0

Columbia University Press
New York Guildford, Surrey

CONTENTS

During the second half of the nineteenth century, the cultivation of phylogenetic trees appeared to be an integral part of biology. In the present century, this form of silviculture has so far been looked on with disfavor, particularly by those experimentalists who did not realize that everything has a history and, since history is the study of the unfinished business of the universe, that history may be highly relevant.

G. E. Hutchinson, 1970,
Foreword to L. Margulis, *Origin of Eucaryotic Cells*
(New Haven: Yale University Press).

PREFACE

Systematists have been interested in the reconstruction of evolutionary history for over a century. Many ideas have been presented on ways to infer patterns of phylogeny, but until the 1950s few explicit methods were developed. Within the past 20 years many new ideas and techniques have been formulated for more precise representations of evolutionary history. Particular focus has been placed on the reconstruction of the branching sequences of phylogeny, and the concepts and methods used in such reconstruction have become known as cladistics.

Because of recent interest in cladistics and the many diverse viewpoints that have been developing, we believed it would be useful to organize a conference that would bring workers with different perspectives together. We also hoped to welcome students (and especially young botanists), so that they might become more familiar with the new concepts and methods and apply them to their own research. The conference was held on March 22–28, 1981, at the University of California, Berkeley. Fifteen contributors (see list on p. xi) each gave one lecture in the morning sessions; their papers are presented here. (Vicki A. Funk, Ronald J. McGinley, David B. Wake, and Allan C. Wilson also gave presentations but did not submit contributions to this volume.) The 28 participants (see list on p. 299) in the afternoon sessions worked on their own data sets, which had been processed before they had arrived using several different computer programs for cladistic analysis. This workshop format allowed for more theoretical discussions in the morning and practical experience in the afternoon, providing a broad introduction to cladistic concepts and methods (see reviews of the workshop in *Syst. Bot.* 6:359–372, 1981, and *Syst. Zool.* 30:491–498, 1981).

We acknowledge the excellent help of many people in bringing this collection of papers to final publication. The National Science Foundation (Grant No. DEB-80-09338) provided the financial support that made the workshop possible and allowed for this published review of cladistics. Richard Arnold, Jeanne Bates, Jim Liebherr, John Sorensen, and Robert Zink helped with the processing of participants' data and with logistic arrangements during the workshop. The cooperation of the Computing Center of the University of California, Berkeley, is also appreciated. We especially thank Richard Jensen for his careful and thorough review of the final manuscript. This book could not have been completed without the constant encouragement, support, and editorial assistance of Vicki Raeburn, Susan Koscielniak, and Amy Fass, to whom we give much credit for the success of this volume.

We hope this book reflects the vitality and exuberance that cladistics has injected into systematic biology. As a community we are not presently undergoing a revolution or conversion, but we are again reevaluating in a healthy fashion old concepts and methods, as we did in the 1960s under the stimulus of phenetics, which has now taken its place among the useful approaches of practicing modern systematists. We believe that cladistics will have a similar positive and lasting effect. We hope that this book is a further step in its development.

T. Duncan
T. F. Stuessy
March 1983

CONTRIBUTORS

Richard A. Arnold
Department of Entomology
University of California
Berkeley, California 94720

Peter D. Ashlock
Department of Entomology
University of Kansas
Lawrence, Kansas 66045

Bernard R. Baum
Biosystematics Research Institute
Agriculture Canada
Ottawa, Ontario, Canada K1A 0C6

Daniel R. Brooks
Department of Zoology
University of British Columbia
Vancouver, British Columbia
Canada V6T 2A9

Jorge V. Crisci
Museo de Ciencias Naturales
Universidad Nacional de La Plata
1900 La Plata, Argentina

Thomas Duncan
Department of Botany
University of California
Berkeley, California 94720

George F. Estabrook
Department of Botany and Herbarium
University of Michigan
Ann Arbor, Michigan 48109

Joseph Felsenstein
Department of Genetics
University of Washington
Seattle, Washington 98195

Walter M. Fitch
Department of Physiological Chemistry
University of Wisconsin
Madison, Wisconsin 53706

David L. Hull
Department of Philosophy
University of Wisconsin
Milwaukee, Wisconsin 53201

Donald R. Kaplan
Department of Botany
University of California
Berkeley, California 94720

Arnold G. Kluge
Department of Ecology and Evolutionary Biology
University of Michigan
Ann Arbor, Michigan 48109

Christopher A. Meacham
Botany Department
University of Georgia
Athens, Georgia 30602

Gareth Nelson
Department of Ichthyology
American Museum of Natural History
New York, New York 10024

Raymond B. Phillips
Department of Botany and Microbiology
University of Oklahoma
Norman, Oklahoma 73019

Tod F. Stuessy
Department of Botany
The Ohio State University
Columbus, Ohio 43210

Warren H. Wagner, Jr.
Department of Botany and Herbarium
University of Michigan
Ann Arbor, Michigan 48109

INTRODUCTION

Since the 1950s, a reexamination of the principles and methods of systematic biology has resulted in much debate. This debate has centered on the theory and methodology of phylogenetic reconstruction and the role such reconstructions should play in the development of classifications. Depending on one's choice, this area of systematic biology is called cladistics or phylogenetic systematics. The first formal statements of these methods and approaches were made by Hennig (1950) and Wagner (1961). Over the last 20 years, discussions of these methods have been along four main lines: the construction of branching diagrams or cladograms as expressions of the evolutionary divergence of organisms; the use of such reconstructions in classification; the development of hypotheses in systematic biology that can be tested and falsified; and the use of estimates of evolutionary relationship as a means for explaining current distributions (historical biogeography).

Other aspects of the phylogenetic history of organisms have also come under scrutiny as part of this debate. The time element, or chronistics, has received the least attention, because of the relative lack of fossil evidence and adequate data on the times of origin of groups. The most significant strides in this area have been made with macromolecular data. The patristic relationship, which is the amount of change that has occurred between any two organisms on a cladogram, is a logical consequence of their positioning according to the various methods of cladogram construction. The significance of this aspect for the development of classifications from cladograms has been debated strongly. The phenetic relationship, which is similarity (and/or difference) shown by using all available characters without consideration of the evolutionary events that produced that similarity (Duncan and Baum 1981), has generated much discussion of its impact on tree reconstruction and attempts at classification.

In this book, a variety of views are presented on these major relationships in phylogenetic reconstruction, as well as comments on many related issues. The controversial nature of cladistics is reflected by disagreements on definitions, concepts, and methods among the contributors to this volume. We leave the reader to evaluate each of the opinions expressed.

Literature Cited

Duncan, T. and B. R. Baum. 1981. Numerical phenetics: Its uses in botanical systematics. *Ann. Rev. Ecol. Syst.* 12:387–404.

Hennig, W. 1950. *Grundzüge einer Theorie der phylogenetischen Systematik.* Berlin: Deutscher Zentralverlag. Translated by D. D. Davis and R. Zangerl, under the title *Phylogenetic Systematics.* Urbana: University of Illinois Press, 1966, rpt. 1979.

Wagner, W. H., Jr. 1961. Problems in the classification of ferns. In *Recent Advances in Botany,* 1:841–844. Montreal: University of Toronto Press.

Cladistics:

Perspectives on the Reconstruction of Evolutionary History

PART I

Philosophical Concerns

INTRODUCTION

Discussion of philosophical issues is commonplace in systematic biology. Systematists, perhaps more than other biologists, have concerned themselves with the philosophical bases of their investigations (Rensch 1968; Hull 1973, 1974; Nelson and Platnick 1981), at least in part as a result of the conceptual and methodological difficulties of understanding the evolutionary process. Discussions have persisted on the nature of species (Mayr 1957; Ghiselin 1974; Hull 1976; Wiley 1980, 1981), the kinds of information needed for making classifications (Crowson 1970) or representing phylogeny (Eldredge and Cracraft 1980; Patterson 1981), the evolutionary meaning of distributions of species and higher taxa (Pielou 1979; Nelson and Rosen 1981), the logical basis of the Linnaean hierarchy (Gregg 1954; Buck and Hull 1966), and so on.

Much philosophical discussion has also ensued over the concepts and methods of cladistics. Virtually every aspect of cladistic analysis has been debated philosophically in detail; much of the discussion can be found in the pages of *Systematic Zoology*. Few, if any, of these issues have been resolved satisfactorily. Serious philosophical differences of opinion still exist on the selection and definition of characters and states, the determination of their polarities, the methods used for cladogram construction, and the relationship between the cladogram and a classification generated from it.

Philosophical issues are addressed to some degree in every paper in this volume. This attests to the importance of philosophical concerns in the present phase of development of cladistics. Three of the papers, however, are almost entirely philosophically oriented. David Hull analyzes the systematic biology community for philosophical views on what we believe we are doing and where we think we are going. Some surprising perspectives are revealed. Arnold Kluge gives viewpoints on the importance of parsimony in science, and more specifically in the reconstruction of phylogeny. Peter Ashlock discusses the different definitions of and philosophical perspectives on monophyly, with a plea for retention of the historical and most common usage.

Literature Cited

Buck, R. C. and D. L. Hull. 1966. The logical structure of the Linnaean hierarchy. *Syst. Zool.* 15:97–111.

Crowson, R. A. 1970. *Classification and Biology.* New York: Atherton Press.

Eldredge, N. and J. Cracraft. 1980. *Phylogenetic Patterns and the Evolutionary Process.* New York: Columbia University Press.

Ghiselin, M. T. 1974. A radical solution to the species problem. *Syst. Zool.* 23:536–544.

Gregg, J. R. 1954. *The Language of Taxonomy.* New York: Columbia University Press.

Hull, D. L. 1973. Contemporary systematic philosophies. *Syst. Zool.* 22:337–400.

——. 1974. *Philosophy of Biological Science.* Englewood Cliffs: Prentice-Hall.

——. 1976. Are species really individuals? *Syst. Zool.* 25:174–191.

Mayr, E., ed. 1957. *The Species Problem.* Amer. Assoc. Adv. Sci. Publ. no. 50.

Nelson, G. and N. Platnick. 1981. *Systematics and Biogeography.* New York: Columbia University Press.

Nelson, G. and D. E. Rosen. 1981. *Vicariance Biogeography: A Critique.* New York: Columbia University Press.

Patterson, C. 1981. Significance of fossils in determining evolutionary relationships. *Ann. Rev. Ecol. Syst.* 12:195–223.

Pielou, E. C. 1979. *Biogeography.* New York: Wiley.

Rensch, B. 1968. *Biophilosophie.* Stuttgart: Gustav Fischer.

Wiley, E. O. 1980. Is the evolutionary species fiction?—A consideration of classes, individuals and historical entities. *Syst. Zool.* 29:76–80.

——. 1981. *Phylogenetics: The Theory and Practice of Phylogenetic Systematics.* New York: Wiley.

Cladistic Theory: Hypotheses That Blur and Grow

David L. Hull

Most empirical investigators simply "do" science. They study stars, continents, species, and hairdressers. A few investigators, however, are engaged in a self-referential activity—investigating empirical investigators. Certain psychologists study the psychological makeup of scientists. They are well aware that anything that they find out must be equally true of them (Mahoney 1979). Sociologists study the social organization characteristic of scientists. Once again, anything that they find about scientists in general should also be true of sociologists, including sociologists of science (Cole 1975). Although historians are not sure that they are "scientists," they are engaged in an empirical activity—chronicling the course of human events, including the history of science. Historians dismiss the histories of science written by scientists as hopelessly biased toward the views that eventually prevailed. Historians propose to treat phlogiston, caloric, and oxygen in equal detail, because these concepts were all equally important in the history of what we now term chemistry. To someone living in Rome, all roads may well appear to lead to Rome, but the historian must be truly a man without a country. Problems arise when historians attempt to write histories of historiography. Self-reference once again rears its ugly head (Hull 1979a).

Most present-day philosophers of science are spared the problem of self-reference because they do not claim to be engaged in an empirical activity. Some claim instead to legislate proper scientific method on the basis of strictly logical considerations. For example, given the logical character of universal laws and deduction, it follows that laws of nature can be falsified but not verified. Other philosophers content themselves with analyzing scientific language. What *do* biologists mean when they use the term "species" or "adaptation"? If philosophers did the sort of empirical research implicit in such an undertaking, they would be engaged in an empirical, albeit linguistic, activity. Too often, however, such activities degenerate into the philosopher's telling us what "we" mean by these terms. This particular circumlocution is a code phrase for the philosopher setting out a proposal for how scientists *should* use the terms being analyzed. Thus far, scientists have rarely found these proposals worth adopting.

To the extent that commentators on science make empirical claims about science, they are committed to presenting evidence for their views and are open to the self-referential gambit. The obvious response to anyone who claims that all generalizations are false is, How about the generalization you just made? Similarly, if generalizations must be falsifiable in order to be meaningful, can that generalization itself be falsified? One response is that such principles as the principle of falsifiability are meant to be metaphysical, not empirical. They are not scientific but metascientific. This response is fine as far as it goes, but it leaves unanswered the question of how metaphysical claims are to be evaluated. Why should one prefer the philosophy of Kuhn (1970) to the philosophy of Feyerabend (1975) or Popper (1959)? Philosophers have been very free about telling scientists how to choose between competing scientific theories. Their efforts at doing the same for themselves have been embarrassingly jumbled.

In my own research I have studied the controversies surrounding pheneticism and cladism in theoretical taxonomy as examples of scientific change. I am hoping to make them test cases for claims that commentators on science make about science. For example, Kuhn and Feyerabend are widely interpreted as saying that scientists who hold different paradigms should not be able to communicate with each other. Their paradigms are incommensurable. In my own research, I have found that scientists frequently do have difficulty in communicating with each other, but this difficulty does not covary universally with their holding different paradigms. Sometimes people who share a paradigm have greater difficulty in communicating with each other than particular scientists who hold different paradigms. The preceding is based on my own perceptions derived from studying the reactions of scientists in one school reading and commenting on the papers of scientists in other schools, e.g., in the refereeing process. Some people are very good at reading a paper in the context of a paradigm which they do not themselves hold; others are not. In addition, most of the scientists whom I have interviewed over the years share my perceptions. It is reassuring that my view of science from the outside coincides with the views of those on the inside. If these views are mistaken, then either they are biased by some factor which influences scientists and commentators on science alike or two errors are being produced by an extremely unfortunate and unlikely confluence of independent factors. But what is a commentator on science to do when his views conflict with those of his subjects?

My study of the pheneticists and cladists over the past 15 years or so has led me to two important conclusions about scientific development: first, that self-interest is an extremely important factor in determining the way that science is conducted, and second, that scientific research programs need not and sometimes do not have any changeless "essence." I do not take these features of science to be accidental. They are inherent to the scientific process. The problem is that when I have informed my subjects of these conclusions, they (surprisingly) tend to agree with the former and deny with flashing eyes and whitened knuckles the latter. I had expected just the opposite reaction.

Most scientists are lucky in that their subjects cannot object to the conclusions drawn about them. However, my subjects can. In this case, who is right, the investigator or the subjects? In this paper I cannot begin to set out the data that I have gathered which have led me to the preceding two conclusions about science. That will have to await a much longer work. However, I will attempt here to explain why the scientists whom I have studied do not share my perceptions about their own research programs. These differences in perception have resulted from my considering the questions, What is phenetic taxonomy and what are its basic principles? and What is cladistic taxonomy and what are its basic principles? The answers given to these questions continue to raise heated disputes, not because the particular answers are "wrong," but because of the underlying assumption that conceptual systems such as pheneticism and cladism can be characterized by eternal, immutable essences. A common assumption which I have discovered in the scientists whom I have studied is that their research program may have changed as far as incidentals are concerned, but not with respect to its basic goals and axioms. I find this conviction peculiar, especially in the case of evolutionary biologists, who have devoted their lives to showing that the parallel conviction for biological species is false.

Altruism in Science

Is science unique or is it like all other human institutions? From my own studies of the recent disputes in theoretical taxonomy, I have come up with the unremarkable conclusion that it is a little bit of both. Many social theorists, on the basis of philosophical considerations, have concluded that by and large people are selfish. They look after themselves and their own first, then worry about humanity at large. Sociobiologists have come to the same conclusion from purely biological considerations. People, like all organisms, should be genetically selfish. Although the social devices for doing so might be complex and obscure, people should devote the vast majority of their efforts to increasing their own inclusive fitness. I happen to think that these global conclusions about the basic selfishness of people in social contexts are not totally accurate. Social institutions can become so involved, and the interconnections so obscure, that people are not infrequently led to behave in ways that are indistinguishable from those commonly thought of as "altruistic." Systems of reciprocal altruism, when practiced by knowledgeable and highly rational people, can easily boil down to reciprocal selfishness. When practiced by people who are easily influenced by appeals to justice and the common good, they can generate genuine altruism.

The question then becomes, What about scientists? Officially, they claim to be altruistic, devoting themselves selflessly to searching after truth for its own sake. One alternative is to conclude that, yes, scientists are unique. While others scramble after the usual rewards in society, scientists are above the fray. Perhaps medical doctors perform twice the number of operations warranted by the circumstances in order to make more money, but scientists would never stoop to such behavior. They are a special breed. A second alternative is that scientists are deceiving themselves and everyone else. They are as interested as everyone else in making money, looking after their families, putting aside a little for a rainy day, and so on. Regardless of how scientists attempt to make it appear, the disinterested quest for truth takes a second seat to their own selfish goals. Elsewhere (Hull 1978a) I have argued for a view that reconciles these two extreme perspectives: Scientists are as self-interested as everyone else but have adopted peculiar goals. The usual rewards of society are secondary to those that are awarded by the scientific community. The primary goal of scientists is to have their ideas accepted by other scientists. In science *use* is the chief form of acceptance. The best thing that one scientist can do for other scientists is to use their work and give them an appropriate citation. The second best thing is simply to use it.

If the mutual cooperation (or exploitation) just described were all there was to science, it would not be especially peculiar. Many other groups are organized in a similar fashion. But in addition, built into the fabric of science is a system of mutual testing. Scientists check their own work, but more importantly, they also check each other's work. Scientific hypotheses must be testable and on occasion actually be tested. If scientific hypotheses existed in isolation, this requirement would be prohibitive. There are simply too many hypotheses. If scientists were required to test hypotheses one by one before they could be incorporated into the body of scientific knowledge, they would still be working on Ptolemaic astronomy. However, scientific hypotheses do not occur in isolation but in inferential systems—scientific theories. Thus, testing one hypothesis in a system serves as an indirect test of the other hypotheses in the system. In point of fact, scientists spend very little time in testing each other's views. They do not have to. The penalties for doing shoddy or deceptive research are so great because it seriously damages the research of other scientists (Zuckerman 1977). Of all self-policing professions, science is genuinely self-policing, because it is in scientists' own self-interest to check the work of their fellow scientists *when it bears on their own research*. Scientists do not test hypotheses at random. They concentrate on those hypotheses that either support or refute their results. The nice thing about the organization of science is that it does not require scientists to go against their own self-interest. The good of the individual usually coincides with the good of the group.

The Nature of Scientific Change

A motto appears beneath a mural in the Zoological Laboratories at the University of Pennsylvania. It reads, Hypotheses That Blur and Grow.[1] A conflict which has characterized evolutionary biology from Darwin and Huxley to the present has been between gradualistic and more saltative (punctuational) forms of evolution. On the one hand, Darwin maintained that traits appear very gradually and change just as imperceptibly in the evolution of a group. Only rarely does a new trait appear fully formed in the space of a single generation. As a result, taxa tend to blur into one another. In speciation a single species becomes subdivided into two large populations, which gradually diverge from each other as minor changes accumulate. In phyletic evolution a single species changes so much through time that organisms from later time-slices might have very little, if anything, in common with organisms from earlier time-slices. On the other hand, saltationists have varied from extreme positions to views so moderate that they blend imperceptibly into gradualism. For example, Schindewolf (1950) maintained that all higher taxa come into existence in the space of a single generation by means of a mechanism similar to Goldschmidt's hopeful monsters (1940). More recently, Eldredge and Gould (1972) have sug-

1. Hampton Carson chose this motto as the title of his contribution to Mayr and Provine's anthology (1980) on the genesis of the synthetic theory of evolution.

gested a microsaltationist view. From an ecological perspective, speciation is a continuous, populational affair, but from the paleontological perspective, it is saltative.

Is scientific development saltative (revolutionary) or gradual (evolutionary)? From my own studies, the answer seems to be, once again, a little bit of both. The change from pre-Darwinian to post-Darwinian views of species is about as abrupt as any change in the history of science. The development of evolutionary theory in the interval has been largely gradualistic, with spurts of activity soon after the turn-of-the-century rise of Mendelian genetics, in the 1920s when evolutionary theory began to be "mathematized," and again in the 1940s with the new synthesis (Mayr and Provine 1980). Currently we seem to be undergoing another flurry of activity. To the extent that speciation is marked by the appearance of a "new" trait universally distributed among the organisms belonging to this new species and absent from all other species, biological species can be treated "essentialistically." Such essences are neither eternal nor immutable, as Aristotle thought, but at least they are discrete. Certainly the classification of plants and animals would be much easier on a saltationist's view of evolution. The same can be said of scientific development. If each new theory contains a hypothesis distinct from those contained in its competitors, then scientific theories might grow, but they would not blur.

I have tried out both of the preceding hypotheses about science on scientists. My suggestions about the role of self-interest in the growth of scientific knowledge seem to strike a responsive chord. Scientists can clearly see this behavior in others and, on a little reflection, in themselves. Perhaps this is not all there is to science, but it is certainly part of the story. The reaction to my suggestion that scientific development is to some extent gradualistic has been near universal and surprisingly vehement denial. Perhaps biological evolution is gradualistic, but not scientific development. Scientific research programs emerge full blown in the writings of their advocates and remain unchanged thereafter, at least in their essentials. For each research program, a set of fundamental tenets exists which *all* and *only* the advocates of this program hold. Anyone who rejects or modifies one of these basic tenets cannot possibly be a member of the "school." The motto implicit in this attitude is that in science nothing less than total allegiance is acceptable. One must hew to the party line or get out.

The preceding view of scientific development may be accurate. I hope not. I find it extremely repugnant. Occasionally, a group of scientists who are attempting to change the direction of science might be fortunate enough to guess right on all issues and never have to change their minds thereafter. Although such happy first guesses are possible, I do not think that they can be as frequent as scientists themselves claim. Besides, I had always thought that one important element of scientific methodology was a mechanism for forcing scientists to change their minds. Perhaps theologians and philosophers doggedly stick with outworn views, but not scientists. Perhaps religious and political groups demand total allegiance, but not groups of scientists. Why then do scientists perceive their own groups in ways antithetical to the very nature of science? I think that the answer to this question is in part sociological. It also depends on how we tend to conceptualize the world.

The first distinction that must be noted is that between groups of scientists and conceptual systems, for example, between the Darwinians and Darwinism. The most common way to define groups and systems is in terms of shared beliefs. The Darwinians are all those scientists who accept the basic tenets of Darwinism. According to Kuhn (1970:176), "A paradigm is what the members of a scientific community share, *and,* conversely, a scientific community consists of men who share a paradigm." If scientific communities are *defined* in terms of shared beliefs, then the preceding claim is a tautology. Kuhn, however, maintains that the two should be and can be defined independently. Scientific communities are to be defined in terms of their social relations, not shared beliefs. Scientific communities, such as the Darwinians, should be defined in terms of cooperation, not agreement. Do all the members of the same scientific community agree with each other, especially about fundamentals? Kuhn claims that yes, they do. The answer that I have discovered in my own research is that no, they do not.

For example, both J. S. Henslow and Richard Owen helped Darwin in his early years. Was either of these men a Darwinian? From a conceptual point of view, Owen was much closer to Darwin than was Henslow. In fact, after the publication of the *Origin,* Owen claimed priority, while Henslow was never able to accept even Darwin's most basic premise that species evolve. Even so, Henslow helped facilitate the reception of Darwin's theory, while Owen worked against it. From a sociological perspective, Henslow was a very important Darwinian, even though he disagreed with Darwin's views. Owen was not. In my own research, I have discovered that scientists can work effectively with other scientists even when they disagree—even over essentials. For example, C. D. Michener was an important member of a group of scientists who, beginning in the late 1950s, investigated ways to make taxonomic judgments more quantitative and explicit—a group that eventually came to be known as the pheneticists, or numerical taxonomists. That Michener belonged to this research group can be documented in a variety of ways, from quantitative measures, such as counting citations, to impressionistic measures, such as asking those concerned. That he never fully shared in the views most commonly and vociferously enunciated by his fellow pheneticists can also be documented. For example, in one of the earliest publications in their emerging research program, Michener and Sokal (1957) explicitly acknowledge differences of opinion—including differences over basics. That Michener's participation in this research program decreased markedly some time prior to 1973 can also be documented. All one needs to do is count references to Michener in Sokal and Sneath (1963) and compare them to references to Michener in Sneath and Sokal (1973). The decline is precipitous.

Defining social groups in terms of social relations is vastly superior to defining them in terms of adherence to shared beliefs. In point of fact, most of the scientists working together to push a particular view of the world will agree with each other, but an important fact about science is that scientists can cooperate even when they disagree. Group membership is extremely important in science. It helps to have fellow workers who will read your papers, comment on them sympathetically, build on your work in their own papers, and so on. It also helps to have sympathetic referees, members on funding panels, and the like. I find such sympathies not in the least shameful. When new research programs are beginning to get underway, they need relatively gentle treatment. Only gradually should more rigorous standards be imposed.

Social groups contain some people who are central, others who are more peripheral. A person who begins as a peripheral member can become more prominent, and vice versa. People can join a group; others leave it. If a group lasts long enough, it is guaranteed to have a total changeover in membership. What is true for social groups in general is true, I think, for groups of scientists. Both Michener and Paul Ehrlich were among the earliest pheneticists. Like Michener, Ehrlich had early reservations, but in Ehrlich's case they were soon overcome, and he became an enthusiastic supporter of the pheneticist viewpoint (Ehrlich 1961). Later he questioned one of the fundamental tenets of phenetic taxonomy—the existence of a general-purpose classification based on some measure of overall similarity. Ehrlich suggested that possibly the most one can hope for is numerous special-purpose classifications (Ehrlich and Ehrlich 1967). Eventually both Michener and Ehrlich ceased active participation in the phenetic school, Michener to work on his bees and to emerge as one of the "enemies" of cladism, Ehrlich to return to his butterflies and to lobby for ecological sanity.

The question now becomes, were both men pheneticists? When did they cease being pheneticists? From the point of view of shared beliefs, Michener was never very much of a pheneticist. Although he thought that this set of beliefs was worth pursuing, he himself accepted very few of the tenets of phenetic taxonomy. Initially, Ehrlich did, but then he came to question one of the most fundamental principles of phenetics. Even though he no longer publishes in taxonomic theory, Ehrlich considers himself a pheneticist, at least in spirit. From a sociological point of view, both Michener and Ehrlich were once pheneticists; at present neither is. As should be obvious by now, the distinction between groups of scientists and the views they may or may not share must be carefully distinguished and kept distinct if confusion is to be avoided. The problem of how to define conceptual systems remains.

Conceptual Development

The easiest way to define a conceptual system is in the same way that many taxonomists define the names of taxa—by means of a particular set of tenets that are severally necessary and jointly sufficient for membership. According to this view, certain tenets are essential, others only incidental. One can have reservations about one or more of the incidental tenets and still be counted as accepting this particular conceptual system. However, one may not have reservations about any of the essential tenets. According to this view, conceptual systems are like territories in a Platonic heaven that people enter and leave as they change their minds. As neat as this way of treating conceptual systems may be, it does not lend itself to conceptual change. If one checks, one discovers that no two "Darwinians" held precisely the same view about evolution. Darwin thought that it was gradual, directed primarily by natural selection, and doubtfully progressive. Huxley agreed but opted for more saltative evolution. Gray held out for some sort of directed evolution, and so on. If "Darwinism" is defined typologically, then the only inhabitant of this Darwinian heaven was Darwin and there is some doubt about him. Conceptual change is possible in this worldview, but it is as jerky as an old silent movie.

Another alternative is to view conceptual systems, as many systematists view the names of taxa, as "cluster concepts." In such systems, certain beliefs are more important than others, but none is necessary. In order to accept a conceptual system using this view, all one has to do is to accept enough of the more important tenets. According to this interpretation, some divergence of opinion is allowed. This second definition of conceptual systems allows conceptual change to be more gradual. Even so, only a certain amount of conceptual change is possible without the abandonment of one conceptual system for another. The biological analog is the division of a single lineage into chronospecies as it undergoes phyletic evolution.

Another alternative is to allow the composition of conceptual systems to change through time *without* subdividing it into separate systems, as long as the system remains cohesive and the change is continuous. According to this interpretation, conceptual systems form historical entities (Ghiselin 1974, 1981; Hull 1976, 1978b; Eldredge and Cracraft 1980; Wiley 1981). This is the alternative that I prefer. According to this view, conceptual systems themselves can evolve. The evolution may be saltative (revolutionary), but it can also be more gradual (evolutionary). Neither mode is ruled out a priori. However, this view also has counterintuitive implications, because views that at one time were competitors to a particular conceptual system can later become part of that system. A proposition can "evolve" into its contradictory.

Although this implication may seem counterintuitive, it accords with the actual evolution of conceptual systems. For example, Darwinism in its early years consisted of a certain cluster of beliefs, not all of them mutually consistent: most versions of "Darwinian" evolution in Darwin's day were progressive, although Darwin's own version was not. In the interim, Darwinism has evolved to exclude this component, as well as others. Two recent challenges to the synthetic theory have been neutralism and punctuational models of speciation. Of course, "Darwinism" at various stages in its development included the acceptance of some neutral characters and various sorts of saltative evolution, but at the time that the neutralists and, subsequently, the punctuationalists wrote, these positions were viewed as "challenges." However, in a recent paper Stebbins and Ayala (1981) argue that both views belong in the "framework of the synthetic theory of evolution." How is this possible? Are there no limits to the cooptation that is possible in scientific development? The answer to this second question is, not much. Critics of the synthetic theory complain of its malleability, its ability to incorporate new discoveries and transmute challenges into new planks for its own platform. The mistake that these critics are making is thinking that this state of affairs is peculiar to the synthetic theory of evolution. On the contrary, it is the result of treating conceptual systems as historical entities. It is equally true of Mendelism, Newtonianism, and every other research program in science. Perhaps phyletic evolution is rare in biological contexts, but it is common in scientific development.

Cladism, Transformed and Untransformed

In my early discussion about the contrast between scientific communities and their conceptual systems, I used pheneticists as an example. It is now the cladists' turn. Several authors have noted that cladism has become "transformed" during the course of its development (Platnick 1979). Have certain tenets remained unchanged in this process or has cladism been completely transformed? If the former, could anyone at the time have predicted which tenets would remain unchanged? If the latter, can a totally transformed cladism still count as the "same" research program? Early on, Hennig (1966) and Brundin (1966) set out what they took to be the principles of "phylogenetics." They included such principles as dichotomy, monophyly, deviation, equal ranking of sister groups, and the extinction of ancestral species at speciation. More than this, both Hennig and Brundin viewed the principles of phylogenetics as being grounded in a particular view of the speciation process. For example, Brundin (1966:23) states quite explicitly:

> It is apparent that the rule of deviation and concepts like plesiomorphy and apomorphy, indeed all the concepts and principles of phylogenetic systematics, are and have to be based on the speciation process, its premises and phylogenetic meaning. The apomorph daughter species is thus identical with the spatially isolated peripheral population of the population genetics which, thanks to successful escape from the rigid system of genetic homeostasis of the mother population, has been able to benefit by the development of new adjustments and to acquire reproductive isolation.

Is Brundin's claim about the theoretical foundations of phylogenetic systematics one of the essential tenets of this school or only a personal idiosyncrasy? Did cladists never hold such views, do they still hold such views, did they once hold such views and have now changed their minds, or what? Answering such questions is not easy, but the task is further complicated by facts about the politics of science. Science is a competitive affair. Scientists set out their views as clearly and comprehensively as possible, emphasizing certain parts, hardly mentioning others. Critics then select certain of these beliefs to challenge. Their choice is not entirely a function of the original authors' intentions. Critics frequently pick what they take to be the weakest tenets to criticize. They also tend to focus on those aspects of other scientists' views that conflict most clearly with their own system of beliefs. The latter must then decide which views to defend. If they spend too much time defending a peripheral belief, it will thereby be elevated to greater importance. If they refuse to respond to a particular criticism, however, their silence is likely to be interpreted as assent. For example, in a letter to Darwin, Charles Lyell (Wilson 1970:467) observed that the "grand argument from absence of mammals & batrachians in Oceanic islands is probably felt to be strong by Owen as he has not ventured to impugn it and therefore declined to bring it into notice."

Did cladists really think that dichotomy was so important? Whether they did or not, this was one of the tenets that elicited the loudest objections (Darlington 1970; Mayr 1974). As a result of both these objections and the subsequent defenses, it became elevated in importance. Thus, any change or abandonment is likely to appear to be a capitulation. Darlington and Mayr hardly approached the cladistic research program from an entirely disinterested, dispassionate perspective. The tenets of cladism that they list as "essential" need not be taken at face value. They are likely to be overly impressed by the ways in which cladistic methods differ from their own, while pheneticists, such as Sokal (1975) and Sneath (1982), are likely to emphasize quite different weaknesses. For example, Sokal is unlikely to complain of the skepticism that cladists such as Nelson (1971, 1972, 1973, 1978) show toward our knowledge of anything about phylogeny except sister-group relations.

For these same reasons, I do not think that the claims made by cladists themselves can always be taken at face value. They too are involved in the ongoing process of science. They too are liable to demote and promote views, depending on later occurrences. At the cutting edge of science, things are not always crystal clear, and degrees of commitment to particular hypotheses change from day to day. In retrospect, one is tempted to remember the doubts about ideas one has since abandoned and forget the periods of strong conviction. Thus, dichotomy does not look as important as it once did. It is a red herring invented by the opponents of cladistic analysis. If I myself had not witnessed and even participated in long "debates" over various principles of cladistic analysis, I would be more prone to adopt the current weighting of these principles.

I do not mean to imply that cladists are especially peculiar in this respect. All scientists reevaluate their beliefs. Nor do I mean to imply that changing one's mind is somehow bad. On the contrary, it is central to science. Neither do I think that objective estimations are impossible to make. For example, any reading of the literature on cladistic analysis shows that monophyly was and continues to be much more important to the cladistic program than dichotomy. I can also understand why scientists tend to downplay any internal disagreements that might crop up. Deemphasis of disagreement among the members of a movement helps to maintain internal cohesion in the group.

Does cladism have an essence? If it does, was it always simply there in the works of Hennig, or did later cladists help to create this essence? In a recent paper on the transformation of cladistics, Platnick (1979:538) argues that, in spite of numerous changes, Hennig's "methods for analyzing data and constructing classifications from them, remain essentially unchanged."

Using this interpretation, "phylogenetics" was from the start essentially a methodological program to develop the logic of constructing branching diagrams and classifications, and nothing else. Of course, Hennig's version included several elements, some of them empirical, as did the early pronouncements of other cladists, but these apparently were incidental to the essence of cladistic analysis.

It is difficult to evaluate such a claim. Certainly Hennig's "phylogenetics" always included a large methodological element, much more than many early critics realized. Many tenets that looked like claims about evolution were not. But I find it difficult to believe that cladistic analysis from the start was nothing but a formal calculus like set theory or matrix algebra. If it is so, I cannot understand the depth of emotion on both sides of the dispute. I can imagine mathematicians and logicians getting all worked up over competing formal systems, but not scientists. If "phylogenetic systematics" from the beginning was really nothing but a methodological program, then many of those involved in this program, as well as their opponents, were confused. If this claim raises the hackles of cladists and anticladists alike, I feel obligated to add that this confusion was extremely productive. I think that neither the proponents nor the opponents of the program could have been roused to expend so much effort on developing the cladist research program if they had fully understood what its essence actually was.

The extent to which cladism has been transformed has yet to be appreciated. Initially, cladism had something to do with evolution, if not the evolutionary process. It dealt with the splitting of branches on phylogenetic trees, the emergence of evolutionary novelties, and so on. In fully transformed cladistics, all that is at issue is nested characters—regardless of whether these characters coincide with speciation events or are themselves evolutionary homologies. At least, this seems to be the drift of recent works in cladistic analysis (Rosen 1978; Patterson 1980; Nelson and Platnick 1981). If I understand these authors, the argument is that either characters can be organized into hierarchic patterns or they cannot. If a hierarchic pattern can be discerned by using a certain set of characters, these characters are to be preferred. These characters may or may not coincide with evolutionary novelties. It would, however, be quite a coincidence if they did not.

I do not know whether the preceding interpretation is accurate. If it is not, I fail to see how cladistic analysis can be simply "methods for analyzing data and constructing classifications from them." If the elements of the analysis are some sort of biological species and the traits evolutionary homologies, then cladistics contains at least two empirical elements. In either case, in order to avoid confusion, two senses of "cladistics" must be distinguished—a general sense, in which cladistics is purely formal, a calculus for generating branching diagrams, and a more special sense, that of the application of this calculus to certain aspects of phylogeny as a biological phenomenon (Hull 1979b). I think that it is cladistics in this special sense that has roused the passions of so many biologists in the recent past, not cladistics in the general sense. Now that this creative confusion has served its purpose, I suggest that it is time to retire it.

The preceding discussion serves a second purpose as well. If the essence of cladism is strictly methodological, the precise nature of this methodology was not immediately clear to everyone concerned. I find the desire to claim that the essence of cladism lay buried in the writings of Hennig (or possibly Brundin) paradoxical. If Hennig set out the essence of cladistic analysis in the beginning, then what credit can later cladists claim for themselves? As I look back over the past decade, its events do not look like a gradual uncovering of a truth that lay buried in Hennig's early writings. On the contrary, I think that later cladists vastly improved on Hennig's system. Perhaps they did not totally transform it, but transform it they did, and they deserve considerable credit for the improvements they made. If someone were to ask me where to go to learn the principles of cladistic analysis, I would not recommend Hennig's *Phylogenetic Systematics*.

I see no reason to interpret the preceding remarks about the development of cladistic analysis as indictments. On the contrary, I think that the features of cladism I have pointed out are characteristics of scientific research programs in general. Science is essentially a creative process. It is not a process in which scientists go from no knowledge to total knowledge but from less knowledge to more knowledge, and this process is never complete. As powerful as the methods of scientific investigation surely are, they never provide final knowledge. If everything essential to the cladistic research program was already at least implicit in the early writings of Hennig, there would have been very little for later workers to do. As it turned out, there was a lot to do. Although there is a strong tendency to claim that all truth resides in the works of some great scientist (was Darwin really a cladist?), I do not think that this belief gives proper credit to later workers. Later cladists did not simply uncover the basic principles of cladistic analysis by a careful reading of Hennig. To some extent, they created those principles.

A Final Turn of the Screw

In studying the principles of biological classification over the past two decades, I have not felt the need to opt for one set or the other. I have never classified an organism, although I have written papers arguing the pros and cons of the principles of classification set out by all three schools. For example, in my earliest publication (Hull 1964), I complained that the principles of evolutionary taxonomy suggested by Simpson (1961) were so weak that only the most variable and impressionistic relation could be established between biological classification and phylogeny (see also Hull 1970). Later I wrote several papers arguing against what I took to be an overly empirical, antitheoretical stance in the work of both the pheneticists (Hull 1967, 1968, 1970) and certain cladists (Hull 1979b, 1980). I now feel these papers to be too one-sided. Of course, scientific classifications cannot be theory-neutral, but scientists are also obligated to make science as "operational" as they can. Operational definitions in the strict sense are inadequate, but the time that scientists spend trying to operationalize their concepts is not wasted.[2] Similarly, it is important to be reminded of the limits of phylogenetic reconstruction. Perhaps claims about past history are falsifiable *in principle,* but *in practice* too often they are not. As is usually the case for philosophers, I have been more interested in in-principle than in-practice issues.

In an indirect way I am forced to choose between the various principles of classification enunciated by the various schools of taxonomy. Perhaps I do not have to classify organisms, but I do have to classify taxonomists. The rules I use to classify the taxonomists I am studying necessarily reveal my own preferences. I do not think that the history of recent disputes in taxonomy can be understood if taxonomists are classified "phenetically." In fact, one of the greatest weaknesses in past histories of science is the classification of scientists in this way. All philosophers and scientists who ever accepted some form of evolution of species, regardless of when and where they lived, are a motley group, a group that played no role whatsoever in the development of evolutionary theory. In conceptual evolution, descent matters. If the history of taxonomy is to make any sense at all, *who* held a view and *where* he got this view is as important as *what* the view actually is. Early on, Nelson (1972, 1973) exhibited extreme doubts about phylogenetic reconstruction. So did the pheneticists. As similar as these doubts may be, in the history of ideas they count as separate and distinct doubts. They exist in different research programs. Like it or not, I am committed to some form of classification in which descent and modification are central.

For anyone following this extended metaphor, the next question natu-

---

2. For example, the *cis-trans* test does not literally ''define'' the gene concept, but it can be used to determine the limits of particular genes in appropriate circumstances.

rally becomes, am I willing to accept paraphyletic groups of taxonomists? Although evolutionary and phenetic taxonomists disagreed on many counts, they agreed that taxa are frequently polythetic (Simpson 1961; Sokal and Sneath 1963; Mayr 1969). Platnick (1979) has argued recently that the reason that taxa seem polythetic is that the traits used are being misidentified. If the principles of cladistic analysis are applied consistently, perfectly nested patterns should materialize. If one insists on retaining traits incompatible with the nested set that one has discovered, then of course the resulting taxa will seem polythetic.

For example, if fins and legs are interpreted as different characters, then the relevant groups appear to be polythetic. However, there is a sense in which fins and legs are the "same" character. They belong to the same transformation series—paired pectoral and pelvic appendages. Neither snakes nor worms have paired pectoral and pelvic appendages. Even so, lacking paired pectoral and pelvic appendages in these two instances are not the same trait. The former must be scored a loss; the latter must be counted as an absence. Just as not all "eyes" are eyes, not all instances of "no limbs" are instances of no limbs. Thus, it is possible for traits to stay "essentially" the same while undergoing extensive modification. Evolving lineages might seem to undergo a total turnover in characteristics, but only if one ignores their origins.

When these same considerations are applied to scientific development, their implications are startling. For example, neither Ed Wiley nor Peter Ashlock accepts the principle of dichotomy, but Wiley's rejection cannot be scored as the same character as Ashlock's rejection. They belong to different transformation series. At one time Wiley accepted the principle of dichotomy. He later came to reject it for a variety of reasons. Ashlock never accepted the principle of dichotomy, for reasons quite different from those of Wiley. Thus, Wiley's rejection belongs in the cladist transformation series (he is the snake), while Ashlock's rejection belongs in the evolutionist transformation series (he is the worm).

At this stage, readers are likely to throw up their hands in dismay. If *this* sort of thing is required for an evolutionary analysis of conceptual development, the results are not worth the effort. I certainly can understand this reaction. Perhaps the usual way of treating the history of ideas is preferable. Certainly it is easier. However, in my own research I plan to follow the logic of the genealogical line of reasoning to its conclusion even if it is an absurdity. If nothing else, I will have shown why others should not follow this same path. However, those taxonomists who think that genealogy is a necessary element in biological classifications should not complain that I have grouped them the way that they group organisms.

Conclusion

What is phenetic taxonomy? Was Paul Ehrlich a pheneticist? Is he still? What is cladistic taxonomy? Was Hennig a cladist? Is Ed Wiley still a cladist? The purpose of this paper has been to show exactly how complicated these apparently straightforward questions actually are and the number of distinctions one must keep in mind to answer them unambiguously. The first is between a scientific group and its conceptual system. If scientific groups are defined in terms of cooperation and not agreement, then it is at least possible for members of the same research group to disagree with each other. Second, if conceptual systems are viewed as historical entities, it is at least possible for them to undergo total change while remaining the same conceptual system. In the process, a particular tenet may evolve into its negation and still belong to the same conceptual transformation series. In this sense, the research program has remained unchanged, and definitions in terms of "essential" traits once again become feasible. I hardly want to praise vagueness and ambiguity. After all, the main purpose of this paper is to decrease both. However, I think that vagueness, at least, is absolutely essential to science; scientific development would be impossible without it (Rosenberg 1975). I think that it is no accident that highly original scientists do not set out their views with absolute clarity when they first enunciate them. Further refinement is always necessary, and sometimes this refinement is so extensive that the original is all but unrecognizable in the product. In order to grow, scientific hypotheses must blur.

Acknowledgments

I owe more than the usual note of thanks to Joel Cracraft, Donald Colless, and Norman Platnick for reading and commenting on an earlier draft of this paper. It was their indignation over that early draft that led me to reassess the ways in which I was individuating phenetics and cladistics as research programs.

Literature Cited

Brundin, L. 1966. Transantarctic relationships and their significance, as evidenced by chironomid midges. *K. svenska Vetensk. Akad. Handl.* 11:1–472.

Cole, S. 1975. The growth of scientific knowledge. In L. A. Coser, ed., *The Idea of Social Structure*, pp. 175–220. New York: Harcourt Brace Jovanovich.

Darlington, P. J. 1970. A practical criticism of Hennig-Brundin "phylogenetic systematics" and antarctic biogeography. *Syst. Zool.* 19:1–18.

Ehrlich, P. 1961. Has the biological species concept outlived its usefulness? *Syst. Zool.* 10:167–176.

Ehrlich, P. and A. H. Ehrlich. 1967. The phenetic relationships of the butterflies. *Syst. Zool.* 16:301–327.

Eldredge, N. and J. Cracraft. 1980. *Phylogenetic Patterns and the Evolutionary Process: Method and Theory in Comparative Biology.* New York: Columbia University Press.

Eldredge, N. and S. J. Gould. 1972. Punctuated equilibria: An alternative to phyletic gradualism. In T. J. M. Schopf, ed., *Models in Paleobiology,* pp. 82–115. San Francisco: Freeman, Cooper.

Feyerabend, P. 1975. *Against Method: Outline of an Anarchist Theory of Knowledge.* London: New Left Books.

Ghiselin, M. 1974. A radical solution to the species problem. *Syst. Zool.* 23:536–544.

Ghiselin, M. 1981. Categories, life, and thinking. *The Behav. & Brain Sci.* 4:269–313.

Goldschmidt, R. 1940. *The Material Basis of Evolution.* New Haven: Yale University Press.

Hennig, W. 1966. *Phylogenetic Systematics.* D. D. Davis and R. Zangerl, trans. Rpt. 1979. Urbana: University of Illinois Press. Originally published as *Grundzüge einer Theorie der phylogenetischen Systematik.* Berlin: Deutscher Zentralverlag.

Hull, D. L. 1964. Consistency and monophyly. *Syst. Zool.* 13:1–11.

——. 1967. Certainty and circularity in evolutionary taxonomy. *Evolution* 21:174–189.

——. 1968. The operational imperative: Sense and nonsense in operationism. *Syst. Zool.* 17:438–459.

——. 1970. Contemporary systematic philosophies. *Ann. Rev. Ecol. Syst.* 1:19–54.

——. 1976. Are species really individuals? *Syst. Zool.* 25:174–191.

——. 1978a. Altruism in science: A sociobiological model of cooperative behavior among scientists. *Anim. Behav.* 26:685–697.

——. 1978b. A matter of individuality. *Phil. Sci.* 45:335–360.

——. 1979a. In defense of presentism. *Hist. & Theory* 18:1–15.

——. 1979b. The limits of cladism. *Syst. Zool.* 28:414–438.

——. 1980. Cladism gets sorted out. *Paleobiology* 6:131–136.

Kuhn, T. *The Structure of Scientific Revolutions.* 2d ed. Chicago: University of Chicago Press.

Mahoney, M. J. 1979. Psychology of the scientist: An evaluative review. *Soc. Stud. Sci.* 9:348–375.

Mayr, E. 1969. *Principles of Systematic Zoology.* New York: McGraw-Hill.

——. 1974. Cladistic analysis or cladistic classification? *Z. Zool. Syst. Evolut.-forsch.* 12:94–128.

Mayr, E. and W. Provine, eds. 1980. *The Evolutionary Synthesis.* Cambridge: Harvard University Press.

Michener, C. D., and R. R. Sokal. 1957. A quantitative approach to a problem in classification. *Evolution* 11:130–162.

Nelson, G. 1971. "Cladism" as a philosophy of classification. *Syst. Zool.* 20:373–376.

———. 1972. Phylogenetic relationship and classification. *Syst. Zool.* 21:227–230.

———. 1973. Classification as an expression of phylogenetic relationship. *Syst. Zool.* 22:344–359.

———. 1978. Ontogeny, phylogeny, paleontology, and the biogenetic law. *Syst. Zool.* 27:324–345.

Nelson, G. and N. Platnick. 1981. *Systematics and Biogeography: Cladistics and Vicariance.* New York: Columbia University Press.

Patterson, C. 1980. Cladistics. *The Biologist* 27:234–240.

Platnick, N. 1979. Philosophy and the transformation of cladistics. *Syst. Zool.* 28:537–546.

Popper, K. R. 1959. *The Logic of Scientific Discovery.* New York: Basic Books.

Rosen, D. 1978. Vicariant patterns and historical explanation in biogeography. *Syst. Zool.* 27:159–188.

Rosenberg, A. 1975. The virtues of vagueness in the languages of science. *Dialogue* 14:281–305.

Schindewolf, O. H. 1950. *Der Zeitfaktor in Geologie und Paläontologie.* Stuttgart: Schweizerbart.

Simpson, G. G. 1961. *Principles of Animal Taxonomy.* New York: Columbia University Press.

Sneath, P. H. A. 1982. Review of *Systematics and Biogeography: Cladistics and Vicariance,* by Gareth Nelson and Norman Platnick. *Syst. Zool.* 31:208–217.

Sneath, P. H. A. and R. R. Sokal. 1973. *Numerical Taxonomy.* San Francisco: Freeman.

Sokal, R. R. 1975. Mayr on cladism—and his critics. *Syst. Zool.* 24:257–262.

Sokal, R. R. and P. H. A. Sneath. 1963. *The Principles of Numerical Taxonomy.* San Francisco: Freeman.

Stebbins, G. L. and F. J. Ayala. 1981. Is a new evolutionary synthesis necessary? *Science* 213:967–971.

Wiley, E. O. 1981. *Phylogenetics: The Theory and Practice of Phylogenetic Systematics.* New York: Wiley.

Wilson, L. G., ed. 1970. *Sir Charles Lyell's Scientific Journals on the Species Question.* New Haven: Yale University Press.

Zuckerman, H. 1977. Deviant behavior and social control in science. *Sage Ann. Rev. Stud. Deviance* 1:87–138.

2

The Relevance of Parsimony to Phylogenetic Inference

Arnold G. Kluge

Parsimony comes from the Latin word *parsimonia,* meaning frugality, and according to current English-language dictionaries, it has two common usages: excessive frugality, and economy of assumption in reasoning. Brown stated that the initial use of parsimony in a scientific context can be traced to Aristotle, whom he quoted as having said, "God and Nature never operate superfluously, but always with the least effort" (1950:135). I believe this to be a purely ontological view of the universe, and that the assumption that the laws of nature are simple is refuted by the history of science (see review, Crisci 1982).

The second, more methodological, usage of parsimony is usually associated with a late medieval scholastic, William of Ockham (c. 1285–1349). His scientific rule of general application is, "We are not allowed to affirm a statement to be true or to maintain that a certain thing exists, unless we are forced to do so either by its self-evidence or by revelation or by experience or by a logical deduction from either a revealed truth or a proposition verified by observation" (Boehner 1957:xx, my trans.). Boehner summarized this principle from various writings by Ockham. The form most often given in Ockham's works is *"Pluralitas non est ponenda sine necessitate* [Plurality is not to be posited without necessity]." Also according to Boehner, "*Frustra fit per plura quod potest fieri per pauciora* [What can be explained by the assumption of fewer things is vainly explained by the assumption of more things]" is infrequently encountered, and the usually cited form, "*Entia non sunt multiplicanda sine necessitate* [Entities must not be multiplied without necessity]," does not appear to have been used by Ockham. According to Crisci (1982:36), John Ponce of Cork probably coined the latter saying in 1639.

Although Ockham was a theologian, he insisted on "evaluations that are severely rational, on distinctions between the necessary and the incidental and differentiation between evidence and degrees of probability" (Vignaux 1979a:505). Ockham used his rule to eliminate the many entities devised by other scholastic philosophers to explain reality, and he used it so frequently and with such sharpness that it became known as "Ockham's razor." The rule of parsimony may have been invoked before Ockham by a contemporary scholastic, Durand de Saint-Pourçain (c. 1270–1334), and a strict scientific application of the principle was also made in the fourteenth century when a French physicist, Nicole d'Oresme, defended the simplest hypothesis of the heavens (Vignaux 1979b).

Wrinch and Jeffries (1921) argued that simple, parsimonious scientific hypotheses are to be preferred over complex hypotheses because they are a priori more probable due to the quality of nature. They went on to say that "scientific practice seems to require the assumption that an inference drawn from a simple scientific law may have a very high probability, often not far from unity" (p. 38). In sharp contrast to their position, Brown asserted that the scientist's attraction to parsimony "does not have its origin in the simplicity of Nature, but in ourselves. We are not able to concentrate attention on more than one train of thought at a given time, and so the simultaneous action of a large number of causes can only be dealt with by endeavouring to calculate what effect would be produced by each separately" (1950:136).

Sober (1981) investigated the use of the principle of parsimony in three areas of science. Phylogenetic inference was not one of these; however, the biological controversy over natural selection at the individual level versus selection at the group level was a source of his information. Sober concluded that parsimony was being applied in other ways than as a deletion rule (Russell 1959). In Sober's opinion, it is often employed as a rule endorsing "induction over previous theoretical claims and explanatory postulates" (1981:155). Bock also emphasized induction; he argued that parsimony has been used by most systematists "as a method by which inductive generalizations are reached from their data. Thus the use of parsimony to generate phylogenetic diagrams with the minimum number of independent origins and parallelisms of characters" (1976:859).

My review of the literature indicates that parsimony has been used in two fundamentally different ways in phylogenetic inference. First, the ontological usage assumes that some quality of nature, say the process of evolution, is economical. I shall refer to this as *evolutionary parsimony* in the remainder of my paper. Crisci's recent attempt (1982) to give this form of parsimony credibility is based on what he called "evolutionary canalization"—more specifically, "selective inertia," "the conservation of organization," and "adaptive modification along the lines of least resistance." I take evolutionary parsimony to mean that descent with modification involves the smallest amount of evolution, viz., no homoplasy (parallelism, convergence, or evolutionary reversal). A relaxed definition of evolutionary parsimony is usually adopted, in which homoplasy is said to be minimal, and at a more proximate causal level this has been interpreted as minimum number of mutations or nucleotide or amino acid changes (Fitch and Margoliash 1967). Second, there is the usage attributable to Ockham, which I designate *methodological parsimony*. This rule urges acceptance of the proposition that best fits all relevant observations and hypotheses. This is the one requiring the smallest number of ad hoc, or a posteriori, assumptions to explain the data.

Evolutionary Parsimony

Sneath (1974) claimed that the assumption that the process of evolution is parsimonious originated with Charles Darwin (no specific reference was given), but I have been unable to confirm his attribution. According to Estabrook (1972), A. W. F. Edwards first suggested the idea of evolutionary parsimony to R. R. Sokal in 1963, and Edwards and Cavalli-Sforza described a method of minimum evolution in that same year. Regardless of the exact time of origin and authorship, there can be no doubt that evolutionary parsimony received its greatest attention in a phylogenetic context in the early 1960s, when the late Joseph Camin made up a phylogeny for a group of hypothetical organisms, subsequently referred to as Caminalcules (Sokal 1966). The true phylogeny, known only to him, was produced by tracing his hypothetical beasts from one piece of paper onto another, and at each tracing preserving all characteristics of the more primitive form, except for a few modifications. Other than gradualism, it is not known what assumptions of evolution Camin followed. Comparing his "true" set of phylogenetic relationships with those guessed by some of his colleagues and the students at the University of Kansas "led him to the observation that those trees which most closely resembled the true cladistics invariably required for their construction the least number of postulated evolutionary steps for the characters studied" (Camin and Sokal 1965:311–312). It was this observation that led to the Camin-Sokal algorithm, a method for deducing branching sequences in phylogeny, and according to Sokal the model assumed, among other things, "that nature is fundamentally parsimonious" (1966:115). Such a proposition about nature, and in particular about long-term evolutionary change, received a lot of criticism (e.g., Rogers, Flemming, and Estabrook 1967). It is easy to falsify a strict evolutionary parsimony assumption for various taxonomic groups and different kinds of data, from molecular to gross morphologic, by simply showing that the minimum-step phylogenetic tree hypothesis almost invariably predicts some incongruent synapomorphies, viz., homoplasy.

Some of the conviction that evolution is parsimonious may stem from our perception that Nature is orderly. For example, cladists conclude non-randomness from congruent synapomorphies (Nelson and Platnick 1981); however, in my opinion such patterns do not provide a basis for accepting even the relaxed view of evolutionary parsimony.

Sober (1975) described a process whereby simplicity judgments can be made about objects and processes, and his approach might be used as well to evaluate the claim that evolution is parsimonious. For example, the simplicity of evolution might be explained in terms of the simplicity of the hypotheses which describe evolutionary processes. I am unaware of any attempts to substantiate the claim that evolution is parsimonious with Sober's approach.

Many of the critics of the Camin-Sokal assumption that evolution is parsimonious were overly zealous and as a result ignored, confused, or denied the appropriateness of parsimony as a methodological rule in phylogenetic inference. For example, Rogers, Flemming, and Estabrook stated that the "correct tree . . . may not be among the most parsimonious ones" (1967:190), and that "the character combinations which are necessary for parsimonious trees may not be biological possibilities" (p. 193). Further, Inger claimed that a most parsimonious hypothesis does not "allow for alteration in the conditions of selection that may lead a population to head first in one genetic direction and then in another" (1967:369). In 1973, Inger concluded that a most parsimonious tree hypothesis "implies that all convergences are equally probable, otherwise a dendrogram with a few more steps might be closer to the true phylogeny than the 'minimum steps' dendrogram" (p. 19). Adding to the misunderstanding, Estabrook (1978) and Funk and Stuessy labeled parsimony methods "guessing procedures" (1978:174), which confuses the ability of existing algorithms to optimize the parsimony criterion with the parsimony rule itself (Farris and Kluge 1979).

Methodological Parsimony

As with evolutionary parsimony, it is difficult to determine the first usage of the methodological rule of parsimony in the context of phylogenetic inference. Kluge and Farris recognized the important difference between the two when they stated that the methodological rule of "parsimony operates by finding a pattern of relationships that is most consistent with the data." They went on to comment that "this may not be a 'biological' reason for choosing between alternative trees; but the principle of tailoring theories to fit known facts is an irreplaceable part of science in general" (1969:7). An even earlier, though less explicit, reference to methodological parsimony can be found in Straus (1949). He endorsed one hominoid phylogenetic hypothesis over another, on the basis of "the principle of parsimony" (p. 219), because it required fewer "difficulties"—presumably ad hoc propositions of homoplasy.

How do competing methods of phylogenetic inference, such as the noncladistic compatibility analysis of Estabrook, Strauch, and Fiala (1977) and the cladistic Wagner Tree algorithm of Farris (1970; see also Farris, Kluge, and Eckardt 1970), square with the methodological rule of parsimony? I believe the compatibility method recognizes only perfect correlations among characters, sets of fully congruent characters, whereas the Wagner Tree method accepts some imperfect correlations, which makes possible a better fit to *all* available data. The former method, usually of necessity, discards data. Scientists should not discard data, for if they do, there is no longer any connection between observation and theory. The theory, whatever it might be called, is no longer scientific (Farris, Kluge, and Mickevich 1982).

The maximum-parsimony algorithm used by molecular biologists is not considered a phylogenetic method, because "there can be no a priori specification of primitive and derived states; these states are assigned only after the nucleotide identities have been statistically sorted on the basis of how often they occur in the descendant sequences" (Goodman et al. 1979:132–133).

The Cladist's Use of Parsimony

My review of phylogenetic systematics indicates that the methodological rule of parsimony is fundamental to cladistics (as are the concepts of synapomorphy and monophyly). The usually encountered form involves a comparison: hypothesis X is preferable to hypothesis Y because X is simpler than Y. Some cladists seem to believe that simplicity and truth are one and the same. The only evidence most cladists accept is character distributions and synapomorphies[1] (Gaffney 1979); from these, with strict ad-

---

1. All homologues are synapomorphies but not all synapomorphies are homologues. Nonhomologous synapomorphies involve independent evolution (convergence, parallelism, and reversal).

herence to the rule of parsimony, conclusions of sister-group relationships are identified.[2] Like the use of methodological rules in general, a cladist's conclusions seem to follow because they are deductive consequences of the chosen rule—methodological parsimony.

Is a cladogram falsifiable? There appears to be considerable difference of opinion, if not confusion, as to the answer (Cartmill 1981), and the problem seems unresolved. If cladograms represent hypotheses about specific historical events among a finite set of objects, then the answer is no (Patterson 1978); the claim is that any monophyletic group (i.e., set of taxa that have a common origin in time and space) cannot be spatiotemporally unrestricted, i.e., not strictly universal. A spatiotemporally restricted class cannot, at the risk of contradiction, be spatiotemporally unrestricted (D. Hull, pers. comm.; Kitts 1977; Ruse 1979; Hull 1982). In contrast to this position, Platnick and Gaffney (1977) concluded that cladograms are strictly universal hypotheses. They reasoned that a cladogram, unlike a phylogenetic tree with its singular statements of common ancestors, does not describe any particular historical event, that a monophyletic group has a temporally unrestricted future, and that the cladistic classification can be reformulated as nonexistence statements (e.g., "there are no members of A that are not more closely related to those of B than they are to those of C"), as can other strictly universal statements. Given these points of view, Platnick and Gaffney went on to state that "cladistic hypotheses can be tested and falsified by the distribution of shared derived characters" (p. 363). Another basis for claiming Popperian falsifiability would be to deny any relationship to evolution, whereupon species and groups would become universals. This position has been adopted by some cladists.

What claim to science is left if a cladogram is not a universal hypothesis? It could be argued (Popper 1980) that testing predictions (or retrodictions) derived from a cladogram saves the scientific character of cladistics. However, as Sparkes (1981) pointed out, it is still possible to protect the hypothesis from which those predictions were made and falsified. While falsifiability is an important factor in any genuine science, it is not the only important one. A good scientific theory is also consilient (Ruse 1979). Perhaps the best argument for the cladists' research program, including use of the methodological rule of parsimony, is its ability to find congruent patterns among different data which other methods do not (Kluge 1983).

2. Cladistics is no longer as conceptually and methodologically homogeneous as it once was, and it is becoming increasingly difficult to generalize about the field. For example, a substantially new position has been taken by what I call "transformed" cladists, those who demand only that groups have characters and be nonoverlapping. Their "chosen rule is not parsimony, but the belief that there is one pattern" in nature (C. Patterson, pers. comm.; see also Patterson 1982:41).

Regardless of whether a cladist uses a computerized algorithm, such as that available for the Wagner Tree method (Farris 1970), or chooses among sets of alternative phylogenetic hypotheses on the basis of the largest number of putative homologues (synapomorphies), in the sense of Hennig's argumentation method (1966), the reason for doing so seems to be the same (see, however, Cartmill 1981). The remainder of this paper reviews several of the bases that cladists might use to justify their use of parsimony.

How and why parsimony is employed in phylogenetic systematics can be illustrated with a simple example (fig. 2.1). For any three supraspecific taxa (A, B, and C) forming a monophyletic group, there are only four (a–d) alternative hypotheses that describe their sister-group relationships. This argument has generality because all hypotheses of monophyletic relationships among terminal taxa can be reduced to the three-taxon case.

While only one of the four hypotheses can be true in the three-taxon example (fig. 2.1), cladists almost always make observations during the course of their studies that support *two or more* of the possible cladistic propositions. In the sense of Hennig (1966), the synapomorphies used to characterize competing monophyletic hypotheses are incongruent. How does the scientist in general, or the phylogeneticist in particular, escape the impasse created by contradictory evidence? The rule of methodological parsimony

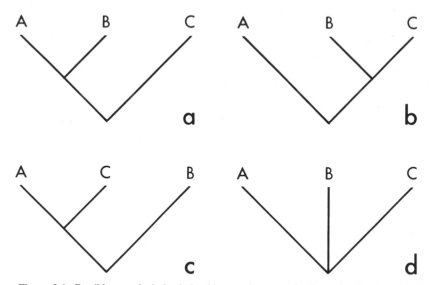

Figure 2.1. Possible genealogical relationships can be conveniently analyzed when only three supraspecific taxa (A, B, and C) are considered, because of the limited number of alternatives (a–d). There are as many as seven possible hypotheses when the taxa are species or subspecific units, because each can be ancestral to the others.

is invoked, and for the phylogeneticist this usually means tentatively accepting for further study the cladistic hypothesis that best fits the data at hand, the one supported by the largest set of congruences, all putative homologues considered.

According to Popper (1968:144–145), choosing the most parsimonious, the simplest, hypothesis is more than a way of avoiding a dead end. "It is a direct corollary of the falsification criterion" for hypothesis testing (Gaffney 1979:98). No scientific hypothesis can be absolutely disproved without a ceteris paribus clause; apparently contradictory singular statements can always be explained with ad hoc hypotheses. In the case of the historical relationships among three taxa (fig. 2.1), there is no way to disprove any of the possible hypotheses, because the phylogeneticist can always "admit the possibility, although perhaps the relative improbability, of convergence" (Farris 1977:397). In this regard, the most parsimonious hypothesis can be defended, because it keeps the investigator from resorting to authoritarianism or a priorism (the auxiliary principle of Hennig 1966; see also Wiley 1975). This justification for invoking parsimony is a particularly critical issue in phylogenetic inference because there appears to be no way to test a hypothesis of homoplasy, independent of the cladogram itself (Cracraft 1972; Gaffney 1979). To be sure, in the three-taxon example (fig. 2.1) one of the cladistic hypotheses is true, and therefore at least one set of the ad hoc hypotheses of homoplasy must also be true. However, aside from discovery of a clerical error in encoding the original observations, I am unaware of a test of homoplasy independent of the phylogenetic hypothesis. Similarly, I view paedomorphosis as an ad hoc hypothesis dependent on a prior assumption of phylogenetic relationships (Kluge 1983).

Popper stated that "it is not so much the number of corroborating instances which determines the degree of corroboration as *the severity of the various tests* to which the hypothesis in question can be, and has been, subjected. But the severity of the tests, in its turn, depends on the *degree of testability,* and thus upon the simplicity of the hypothesis" (1968:267). Simpler hypotheses are more falsifiable, since they permit fewer exceptional clauses, and in the same regard simpler hypotheses are a priori less probable than complex hypotheses. The simplest hypothesis compatible with the data is the more falsifiable and improbable hypothesis. It represents a more severe test than the complex hypothesis. "Thus, to the extent that simpler hypotheses are more falsifiable, simplicity is a characteristically scientific aim" (Beatty and Fink 1979:644).

Popper (1968) also encouraged the selection of parsimonious hypotheses because of their greater informativeness. According to Beatty and Fink, he began with the observation that "the least falsifiable, most probable hypothesis is that which merely summarizes the data" (1979:644). Alternatively, the most falsifiable, least probable hypothesis is the one containing the most information beyond that present in the data set. Numerous counterexamples to high probability/low falsifiability and low probability/high falsifiability have been constructed (Beatty and Fink 1979; see also Salmon 1966), and Sober (1975) sought an alternative to Popper's probabilistic interpretation of simplicity.

Sober (1975) also used informativeness, but in a way that differed from Popper's concept. Sober's measure of "informativeness of a hypothesis is the measurable extent to which the hypothesis alone answers questions about individuals in its domain" (Beatty and Fink 1979:645). "The more informative our knowledge claims are about the individuals in our environment, the less we need to find out about the special details of an arbitrary individual before we can say what its properties are" (Sober 1975:3). More specifically, "the informativeness—hence simplicity—of a hypothesis H with regard to question Q (about an individual a_i in the domain of H) is inversely proportional to the amount of extra information which must be added to H to answer Q" (Beatty and Fink 1979:645). In terms of phylogenetic inference, the cladistic hypothesis requiring fewer evolutionary assumptions to account for a particular data set *will* be more informative than one requiring more assumptions. This is because the simplest cladogram requires less extra information than a more complex one to be able to answer a question about the explanation of a particular data set (E. Sober, pers. comm.).

Similarly, Farris (1979, 1980) made a strong case for parsimony through its relationship to descriptive information content. He noted that "the classification that allows data to be described in the smallest number of symbols is therefore that corresponding to a most parsimonious tree" (1979:508). Regardless of their historical accuracy, such trees are the most informative because they correspond to the classifications with the most efficient diagnoses.

Is there any reason to believe that simpler, more parsimonious hypotheses are more likely to be predictive? According to Beatty and Fink, "if 'predictive informativeness' (about previously unstudied characters) were synonymous with 'informativeness' in Sober's sense" (1979:650), there would be an additional reason to accept parsimony as a goal of phylogenetic inference. Another possible reason lies in the relationship between monophyletic groups and natural kinds (J. Farris, pers. comm.).

The twin concepts of natural kinds and the natural system of classification were prominent in the writings of nineteenth-century philosophers of science, especially William Whewell, John Stuart Mill, and William Stanley Jevons. Natural kinds were, and still are, viewed as groupings that correspond to reality, and the natural system of classification of those kinds as the basis for formulating scientific laws (Hull 1982). Their importance was succinctly underscored by Jevons' belief that "science can extend only so far as the power of accurate classification extends" (1874:421).

According to some of the nineteenth-century philosophers (e.g., see Whewell 1847), natural kinds, like natural laws, are eternal and immutable, and from that perspective species cannot be examples of natural kinds (Ghiselin 1974; Hull 1976, 1978). Mill (1862), among others of the time, seems to have held the opposite opinion (e.g., his analogy of snub-nosed humans). The contradiction of a species as a natural kind is avoided if species, as well as cladistic classifications, are conceived of as natural individuals. "They are natural individuals because they function *in* natural processes" (D. Hull, pers. comm.). As individuals, they serve as entities which instantiate their natural kind.

The relationship between predictiveness and natural kinds may be judged from Whewell's XIV Aphorism Concerning Science: "*The Consilience of Inductions* takes place when an Induction, obtained from one class of facts, coincides with an Induction, obtained from another different class. This Consilience is a test of the truth of the Theory in which it occurs" (1847, 2:469; see also Ruse 1979). The emphasis in Whewell's Aphorism is on one class of facts predicting another, *different* class and thereby being consilient. For example, a set of congruent morphological synapomorphies for hominoids predicts certain parasite relationships, whereas most molecular synapomorphies do not. From this I argued (Kluge 1983) that the congruence between morphological and host-parasite data is a case of consilience and went on to conclude that such congruent synapomorphies are comprehensible only when the groups they diagnose are real, viz., the products of a natural process. In biology, as convincingly set forth by Charles Darwin (1871), natural kinds are groups that evolve via descent with modification and the natural system is strictly genealogical.[3] The reason that genealogically arranged groups are predictive is that they represent separate breeding communities and the natural distributions of an indefinitely large number of features correspond to the boundaries of communities of

3. Darwin stated that "naturalists have long felt a profound conviction that there is a natural system. This system, it is now generally admitted, must be, as far as possible, genealogical in arrangement—that is, the co-descendants of the same form must be kept together in one group, separate from the co-descendants of any other form; but if the parent-forms are related, so will their descendants, and the two groups together will form a larger group" (1871, 1:181).

descent. Assuming a strictly genealogical natural system, monophyletic groups are natural kinds in just the same way that species are, because the process that produces both is the same (J. Farris, pers. comm.).

An additional reason for choosing the most parsimonious hypothesis may be founded on the claim that it is more likely to be true (as a hypothesis) than an alternative supported by fewer congruent synapomorphies (see fig. 2.1). This involves the assumption that there is no a priori reason to expect homoplasious characters, as a group, to be congruent. Further, such false synapomorphies are expected to be uncorrelated with true synapomorphies (homologues) and uncorrelated among themselves as well. The processes of paedomorphosis and separate clades independently evolving into the same adaptive zone may be potential tests of this proposition; however, both assume some cladistic hypothesis.

Summary

The relevance of methodological parsimony to phylogenetic systematics may be founded on the following arguments:

1. The choice of the simplest hypothesis provides a rationale for escaping the impasse created by contradictory evidence.

2. Choice of the simplest hypothesis keeps the investigator from resorting to authoritarianism and a priorism.

3. The simplest hypothesis is subject to the most severe test.

4. The simplest hypothesis is the most informative (sensu Sober).

5. The simplest hypothesis provides the most efficient diagnoses.

6. Simpler hypotheses are more predictive and lead to the identification of natural kinds.

7. Under certain conditions, the simplest hypothesis may be preferred because it is more likely to be true (as a hypothesis).

Acknowledgments

My brief review of parsimony could not have been completed without access to unpublished manuscripts and special insights provided by colleagues. I am grateful for the many contributions made by John Beatty, James S. Farris, Bill Fink, Sara Fink, David Hull, Colin Patterson, and Elliot Sober. I am indebted to David Hull for piquing my conscience.

Literature Cited

Beatty, J. and W. L. Fink. 1979. Review of *Simplicity*, by Elliot Sober. *Syst. Zool.* 28:643–651.

Bock, W. 1976. Foundations and methods of evolutionary classification. In M. K. Hecht, P. C. Goody, and B. M. Hecht, eds., *Major Patterns of Vertebrate Evolution*, pp. 851–895. New York: Plenum Press.

Boehner, P. 1957. *Ockham—Philosophical Writings*. Indianapolis: Bobbs-Merrill.

Brady, R. H. 1983. Parsimony, hierarchy, and biological implications. In N. I. Platnick and V. A. Funk, eds., *Advances in Cladistics, Vol. 2: Proceedings of the Second Meeting of the Willi Hennig Society*, pp. 49–60. New York: Columbia University Press.

Brown, G. B. 1950. *Science: Its Method and Its Philosophy*. London: George Allen and Unwin.

Camin, J. H. and R. R. Sokal. 1965. A method for deducing branching sequences in phylogeny. *Evolution* 19:311–326.

Cartmill, M. 1981. Hypothesis testing and phylogenetic reconstruction. *Z. zool. Syst. Evolut.-forsch.* 19:73–96.

Cracraft, J. 1972. The relationships of the higher taxa of birds: Problems in phylogenetic reasoning. *Condor* 74:379–392.

Crisci, J. V. 1982. Parsimony in evolutionary theory: Law or methodological prescription? *J. Theor. Biol.* 97:35–41.

Darwin, C. 1871. *The Descent of Man, and Selection in Relation to Sex*. 2 vols. New York: Appleton.

Edwards, A. W. F. and L. L. Cavalli-Sforza. 1963. The reconstruction of evolution. *Ann. Human Genet.* 27:104–105.

Estabrook, G. F. 1972. Cladistic methodology: A discussion of the theoretical basis for the induction of evolutionary history. *Ann. Rev. Ecol. Syst.* 3:427–456.

——. 1978. Some concepts for the estimation of evolutionary relationships in systematic botany. *Syst. Bot.* 3:146–158.

Estabrook, G. F., J. G. Strauch, and K. L. Fiala. 1977. An application of compatibility analysis to the Blackith's data on orthopteroid insects. *Syst. Zool.* 26:269–276.

Farris, J. S. 1970. Methods for computing Wagner trees. *Syst. Zool.* 19:83–92.

——. 1977. Commentary. *Syst. Zool.* 22:396–398.

——. 1979. The information content of the phylogenetic system. *Syst. Zool.* 28:483–519.

——. 1980. The efficient diagnoses of the phylogenetic system. *Syst. Zool.* 29:386–401.

——. 1983. The logical basis of phylogenetic systematics. In N. I. Platnick and V. A. Funk, eds., *Advances in Cladistics, Vol. 2: Proceedings of the Second Meeting of the Willi Hennig Society*, pp. 7–36. New York: Columbia University Press.

Farris, J. S. and A. G. Kluge. 1979. A botanical clique. Review of *Cladistics and Plant Systematics*, by T. Stuessy and G. F. Estabrook. *Syst. Zool.* 28:400–411.

Farris, J. S., A. G. Kluge, and M. J. Eckardt. 1970. A numerical approach to phylogenetic systematics. *Syst. Zool.* 19:172–189.

Farris, J. S., A. G. Kluge, and M. F. Mickevich. 1982. Phylogenetic analysis, the monothetic group method, and myobatrachid frogs. *Syst. Zool.* 31:317–327.

Felsenstein, J. 1983. Parsimony in systematics: Biological and statistical issues. *Ann. Rev. Ecol. Syst.* 14:313–333.

Fitch, W. M. and E. Margoliash. 1967. Construction of phylogenetic trees. *Science* 155:279–284.

Funk, V. A. and T. F. Stuessy. 1978. Cladistics for the practicing plant taxonomist. *Syst. Bot.* 3:159–178.

Gaffney, E. S. 1979. An introduction to the logic of phylogeny reconstruction. In J. Cracraft and N. Eldredge, eds., *Phylogenetic Analysis and Paleontology,* pp. 79–111. New York: Columbia University Press.

Ghiselin, M. 1974. A radical solution to the species problem. *Syst. Zool.* 23:536–544.

Goodman, M., J. Czelusniak, G. W. Moore, A. E. Romero-Herrera, and G. Matsuda. 1979. Fitting the gene lineage into its species lineage, a parsimony strategy illustrated by cladograms constructed from globin sequences. *Syst. Zool.* 28:132–163.

Hennig, W. 1966. *Phylogenetic Systematics.* D. D. Davis and R. Zangerl, trans. Rpt. 1979. Urbana: University of Illinois Press.

Hull, D. 1976. Are species really individuals? *Syst. Zool.* 25:174–191.

——. 1978. A matter of individuality. *Phil. Sci.* 45:335–360.

——. 1982. The principles of biological classification: The use and abuse of philosophy. In P. Asquith and R. Giere, eds., *PSA 1978,* 2:130–153. East Lansing, Mich.: Philosophy of Science Association.

Inger, R. F. 1967. The development of a phylogeny of frogs. *Evolution* 21:369–384.

——. 1973. Numerical taxonomy. *Caldasia* 11:7–28.

Jevons, W. S. 1874. *The Principles of Science: A Treatise on Logic and Scientific Method.* Vol. 2. London: Macmillan.

Johnson, R. 1982. Parsimony principles in phylogenetic systematics: A critical re-appraisal. *Evol. Theory* 6:79–90.

Kitts, D. B. 1977. Karl Popper, verifiability, and systematic zoology. *Syst. Zool.* 26:185–194.

Kluge, A. G. 1983. Cladistics and the classification of the great apes. In R. L. Ciochon and R. S. Corruccini, eds., *New Interpretations of Ape and Human Ancestry,* pp. 151–177. New York: Plenum.

Kluge, A. G. and J. S. Farris. 1969. Quantitative phyletics and the evolution of anurans. *Syst. Zool.* 18:1–32.

Mill, J. S. 1862. *System of Logic, Ratiocinative and Inductive, Being a Connected View of the Principles of Evidence, and the Methods of Scientific Investigation.* 5th ed. London: Parker, Son, and Brown.

Nelson, G. and N. Platnick. 1981. *Systematics and Biogeography.* New York: Columbia University Press.

Panchen, A. L. 1982. The use of parsimony in testing phylogenetic hypotheses. *Zool. J. Linn. Soc.* 74:305–328.

Patterson, C. 1978. Verifiability in systematics. *Syst. Zool.* 27:218–222.

———. 1982. Morphological characters and homology. In K. A. Joysey and A. E. Friday, eds., *Problems of Phylogenetic Reconstruction*, pp. 21–74. New York: Academic Press.

Platnick, N. I. and E. S. Gaffney. 1977. Systematics: A Popperian perspective. *Syst. Zool.* 26:360–365.

Popper, K. R. 1968. *The Logic of Scientific Discovery*. 2d ed. New York: Harper and Row.

———. 1980. Evolution. *New Scientist* 87:611.

Rogers, D. J., H. S. Flemming, and G. Estabrook. 1967. Use of computers in studies of taxonomy and evolution. *Evol. Biol.* 1:169–196.

Ruse, M. 1979. Falsifiability, consilience, and systematics. *Syst. Zool.* 28:530–536.

Russell, B. 1959. *My Philosophical Development*. London: George Allen and Unwin.

Salmon, W. 1966. *The Foundations of Scientific Inference*. Pittsburgh: University of Pittsburgh Press.

Sneath, P. H. A. 1974. Phylogeny of microorganisms. *Symposia Soc. Gen. Microbiol.* 24:1–39.

Sober, E. 1975. *Simplicity*. Oxford: Clarendon Press.

———. 1981. The principle of parsimony. *Brit. J. Philos. Sci.* 32:145–156.

———. 1983a. Parsimony methods in systematics. In N. I. Platnick and V. A. Funk, eds., *Advances in Cladistics, Vol. 2: Proceedings of the Second Meeting of the Willi Hennig Society*, pp. 37–47. New York: Columbia University Press.

———. 1983b. Parsimony in systematics: Philosophical issues. *Ann. Rev. Ecol. Syst.* 14:335–357.

Sokal, R. R. 1966. Numerical taxonomy. *Sci. Amer.* 215:106–116.

Sparkes, J. 1981. What is this thing called science? *New Scientist* 89:156–158.

Straus, W. L., Jr. 1949. The riddle of man's ancestry. *Quart. Rev. Biol.* 24:200–223.

Vignaux, P. D. 1979a. Ockham, William of. In *Macropaedia*, vol. 13, pp. 504–506. Chicago: Encyclopaedia Britannica, Inc.

———. 1979b. Ockham's razor. In *Micropaedia*, vol. 7, pp. 475–476. Chicago: Encyclopaedia Britannica, Inc.

Whewell, W. 1847. *The Philosophy of the Inductive Sciences, Founded Upon Their History*. 2 vols. London: Parker.

Wiley, E. O. 1975. Karl R. Popper, systematics, and classification: A reply to Walter Bock and other evolutionary taxonomists. *Syst. Zool.* 24:233–243.

———. 1981. *Phylogenetics*. New York: John Wiley and Sons.

Wrinch, D., and H. Jeffries. 1921. On certain fundamental principles of scientific inquiry. *Philos. Mag.* 42:369–390.

Addendum

The numerous articles on parsimony and phylogenetic systematics that have appeared since I submitted this manuscript on June 8, 1981, are ample testimony to the importance and controversial nature of the topic (see, for example, Brady 1983; Felsenstein 1983; Johnson 1982; Panchen 1982; Sober 1983a, 1983b; Wiley 1981). I believe Farris' in-depth treatment (1983) is an especially important contribution, and I urge that it be read; it covers several topics not considered herein.

3

Monophyly: Its Meaning and Importance

PETER D. ASHLOCK

The idea that taxa should be monophyletic had its beginning in the writings of Charles Darwin. He stressed that "all true classification is genealogical" (Darwin 1959:656) but that "genealogy by itself does not give the classification" (F. Darwin 1888:247). Expanding the idea, he wrote, "The arrangement of groups within each class, in due subordination and relation to other groups, though allied in the same degree in blood to their common progenitor, may differ greatly, being due to the different degrees of modification which they have undergone; and this is expressed by forms being ranked under different genera, families, sections, or orders" (Darwin 1959:656). In his classification of the barnacles, Darwin consistently determined rank by degree of divergence, and many of his groups have been shown to have evolved from within other groups (Ghiselin and Jaffe 1973; Mayr 1974).

One of Darwin's most ardent supporters was the highly prolific Ernst Haeckel, about whom Darwin wrote, "Professor Häckel in his 'Generelle Morphologie' and in other works, has recently brought his great knowledge and abilities to bear on what he calls phylogeny, or the lines of descent of all organic beings. . . . He has thus boldly made a great beginning and shows us how classification will in the future be treated" (Darwin 1959:676). Haeckel in the early 1860s coined several terms for some of Darwin's concepts, and Haeckel's first use in English of the terms phylogeny, monophyletic, and polyphyletic appeared in 1874.

In his 1874 contribution, Haeckel argued against the "type theory" of his contemporaries, who pointed to the existence of seven distinct and isolated types: Protozoa, Coelenterata, Vermes, Mollusca, Arthropoda, Echinodermata, and Vertebrata. Haeckel cites an adversary named Hopkins (no

Contribution number 1800 from the Department of Entomology, University of Kansas.

references given), who, calling these types the Kepler's Laws of the animal kingdom, saw in them the "most brilliant confutation of the Darwinian heresy" (Haeckel 1874:240–241). This multiple origin of the animal kingdom Haeckel called polyphyletic descent (p. 241).

Haeckel argued strongly (pp. 241–243) for a single or monophyletic origin of the animal kingdom. He cites his own work and that of contemporaries, and writes, "Through this demonstrable connection of the phyla it will appear that the whole of the members of the animal kingdom can be placed in a near alliance, whereby the ground is got ready for a monophyletic system" (pp. 242–243).

Until the middle of the twentieth century, systematists generally used *monophyletic* and *polyphyletic* to refer to single and multiple origins of groups of organisms. Then various theoreticians, including Simpson (1961), discussed the ambiguity of the terms and the need for more rigorous definitions. Simpson's definition of monophyletic, however, did not exclude clearly polyphyletic groups (see Tuomikoski 1967; Ashlock 1979).

Hennig (1966) added rigor to the definition by requiring that a monophyletic group consist of all descendants of its stem ancestor. Moreover, only such monophyletic groups should be recognized formally in classifications. He coined a companion term, paraphyletic, for groups derived from a single stem species but not containing all descendants of that ancestor. This narrow concept of monophyletic groups substantially differed from the traditional concept, and actual application of Hennig's definition to the practice of systematics would produce gross changes in classification.

Tuomikoski (1967) continued the discussion of monophyly in his critical review of some of Hennig's principles. He suggested that paraphyletic groups be tolerated in classifications. He concluded, "Once the phylogeny of a group is studied as critically as possible, the most certain results illustrated by the tree diagram, and the classification of the group revised to the extent that is found necessary and practical, any remaining incongruence between phylogeny and classification will be of secondary importance and no longer hinder the understanding of the evolution of the group" (p. 147).

In 1971 I coined a new term, holophyletic, for Hennig's all-inclusive group, and proposed the following set of definitions (given here in slightly modified form) for monophyly and associated terms: A monophyletic group is one whose most recent common ancestor is a member of that group. A monophyletic group may be either holophyletic (containing all descendants of the most recent common ancestor of that group; monophyly sensu Hennig) or paraphyletic (not containing all descendants of the most recent common ancestor of the group). In contrast with the monophyletic group is the polyphyletic group, one whose most recent common ancestor is not a member of that group. The 1971 version required that the common ancestor be a "cladistic" member of the group, which created some confusion. After thought, I have concluded that removing the word does not alter the definitions, and I have done so.

As in all definitions of these terms, the common ancestor may be inferred, and the definitions are not based on the use of actual characters; but as in the use of all such definitions, evidence for relationship and group membership must come from an analysis of characters. Mayr (1974), Ross (1974), Bock (1977), and Charig (1982), among others, have all strongly encouraged use of these terms defined in this manner.

Nelson (1971) responded to my 1971 paper by redefining the terms exclusively by reference to cladistic branching. He defined monophyletic groups as "complete sister group systems" and found nonmonophyletic groups to be of two kinds: paraphyletic, a sister-group system incomplete because of the omission of one species or independent monophyletic species-group; and polyphyletic, a sister group incomplete because of the omission of two such species or groups. These definitions not only accept Hennig's nontraditional definition of monophyly but also divide Hennig's paraphyly in two parts and apply Haeckel's term polyphyly to one of them (Ashlock 1972). Nelson's two sorts of "non-monophyletic" groups do not address the question of single or multiple origin of such groups, and Haeckel's useful concept of polyphyly is ignored.

Farris (1974), noting that Nelson had altered Hennig's concept of paraphyletic groups, provided another set of definitions: "The group is said to be monophyletic if its group membership character appears uniquely derived and unreversed . . . paraphyletic if its group membership character appears uniquely derived, but reversed . . . [and] polyphyletic if its group membership character appears non-uniquely derived" (p. 554). In an explication of Farris' definitions, Platnick helpfully simplified "group membership character" to "group membership" (1977:198). After reflection, it becomes clear that requiring "uniquely derived group membership" is the same as including "the most recent common ancestor as a member of the group." In Farris' analysis, "reversed" describes a group from which some descendant has been excluded, so "group membership reversed or not" refers to whether "all descendants are included or not." Farris' definitions, then, are rewordings of the same principles I used in 1971 and 1972 (see Ashlock 1979).

Wiley (1979) had difficulty with my 1971 definitions when he tried to apply them to a group consisting of two sister species and their immediately ancestral species, whose ancestor also gave rise to another species, excluded from Wiley's group. The earlier ancestor cannot, he says, be a member of the original group of three, for it is also the ancestor of an excluded species. Consequently, he concludes that the original group of three is polyphyletic by my definitions.

A similar line of reasoning was applied to the question of the monophyly of a single taxon by Platnick (1976) when he asked, "Are monotypic genera possible?" Wiley in 1977 argued that monotypic taxa (i.e., single species) are monophyletic a priori and suggested that if Platnick's reasoning were followed, no taxon of more than one species could be monophyletic either. The *reductio ad absurdum* is that if the ancestor of the ancestor must be considered, then the ancestor unto the *n*th generation must also be considered. Thus, the only possible monophyletic group is all of creation.

Although in 1979 Wiley termed this "infinite series" only an "apparent paradox" (p. 313), the ancestor-of-the-ancestor reasoning that produced it would apply to all recent definitions of monophyly or holophyly, including those used by cladists, and these all require that the ancestor be included in the group. None requires the ancestor of the ancestor to be considered. There is no paradox. The ancestor of a monotypic genus is the only included species of that genus. The ancestor of a larger monophyletic group is the included stem ancestor, and it is irrelevant (Wiley 1979 to the contrary) whether that stem ancestor is inferred or is represented by a fossil in hand.

All current definitions of monophyly and associated terms use, either singly or in combination, two criteria: existence of a single and immediate ancestor or not; and inclusion of all descendants of that common ancestor or not. These criteria, variously worded, are included in the definitions of Hennig (1966), Ashlock (1971), Farris (1974), and Platnick (1977).

J. S. Farris (pers. comm.) argued for priority of the Hennig definition of monophyly because it is like Haeckel's. Taken by itself, Haeckel's wording (1874)—roughly, a monophyletic concept of the animal kingdom as a single root-form and all of its descendants—is much like Hennig's definition (1966) of monophyly. Moreover, in the context in which Haeckel's definition was proposed, the origin of the whole animal kingdom, a Hennig-like definition is entirely appropriate. But the traditional meaning of a word depends not only on the very general definition it may first have received, but also on subsequent usage. In 1874 and 1894, Haeckel refers to his monophyletic concept of the animal kingdom vs. other workers' polyphyletic concepts and to his monophyletic system, which connects all of the phyla. He demonstrates the connection on phylogenetic trees, often with major characters for various collective groups listed in an accompanying chart. One such tree, headed "Monophyletic stem-structure of the animal kingdom" (1874:247), is somewhat ambiguous by modern standards, but at least three paraphyletic groups are clearly portrayed on that tree. In his 1894 work, *Systematic Phylogeny of Protists and Plants,* the several trees throughout the book are all headed "Monophyletischer Stammbaum," monophyletic phylogeny, of the group concerned. In the phylogeny of the Infusoria, one finds a paraphyletic Flagellata from within which evolve a holophyletic Ciliata, a holophyletic Suctoria, and a holophyletic Metazoa. Haeckel's monophyletic Infusoria, then, is paraphyletic. Of twelve monophyletic phylogenies in the 1894 work, which treat nearly 400 higher taxa, over 40 percent are clearly paraphyletic. I can find no evidence that Haeckel or any of his successors up to 1950 used *monophyletic* or *polyphyletic* in any sense other than simply single or multiple origins of taxa.

The only difference between my own treatment of the terms and those of the cladists is that I use four, including holophyly, and many cladists use three, excluding holophyly. I use Haeckel's traditional distinction of single or multiple origin for the contrasting pair *monophyletic* and *polyphyletic.* For those groups of a single origin, I use another contrasting pair, *holophyletic* and *paraphyletic,* for complete or incomplete inclusiveness. The cladists have one term, *polyphyletic,* for lack of single origin; another, *paraphyletic,* for lack of complete inclusiveness; and a third, *monophyletic,* that combines single origin and complete inclusiveness.

Another criterion for monophyly seems self-evident: complete interconnectedness. A group cannot be monophyletic by any definition, even if it has a single stem ancestor, if one or more lineages have left the circumscribed monophyletic group and then apparently rejoin it because of convergent characters through a derived lineage. The need for total connectedness within monophyletic groups was emphasized by Duncan (1980), who uses the term "convex," introduced by Estabrook (1978). I accept the criterion that traditional phyletic groups must be convex, i.e., totally interconnected within the monophyletic group.

Hull wrote, "The debate between cladists and evolutionists over the 'proper' definition of 'monophyly' is not an idle dispute over terminology . . ." (1980:134). He refers to cladists' objection to the idea that a higher taxon as a unit can be thought of as ancestor to another higher taxon; such claims are in conflict with the cladists' requirement of strict monophyly (holophyly) and with the requirements of common sense as well.

Whenever systematists have, in the past, stated that one higher taxon evolved from another, they either had some peculiar, perhaps teleological, idea about the evolutionary process or were using sloppy language. No sensible modern systematist would state that, say, the whole of the reptiles gave rise to the whole of the birds. The sloppy language that gave rise to this misconception should be corrected to: one monophyletic grade may be the sister group of *part* of another monophyletic grade.

The importance of the Hennigian revolution to an evolutionary systematist is in the analytical concepts that were developed and refined. These are that phyletic branching is ultimately derived from speciation, that monophyletic grades may be holophyletic or paraphyletic, and that the way to determine the sequence of phyletic branching is to find nested sets of holophyletic groups each of which is supported by synapomorphies.

These principles are obvious now, but many ideas are not in the least obvious until they have been articulated. Lacking these concepts, phylogeneticists back to Darwin and Haeckel had no clear idea of how to establish a defensible sequence of branching. Too often, their sequences were either devoid of characters, reflecting what might be called a phenetic rather than an evolutionary concept of relationship, or were an indiscriminate mix of primitive and derived characters. Though one can find older work in which Hennig's analytical concepts were grasped and used sporadically, it is the consistent use of the whole set of interrelated concepts that is necessary for a rigorous analysis.

A modern, properly constructed phylogenetic diagram expresses better than any other known technique cladogenesis, anagenesis, and, when appropriate, the chronology of development of a group of living things. No formal classification can express all of these things simultaneously. A diagram, already limited to two dimensions, cannot be expressed better in a one-dimensional (even though hierarchical) list of names. It is what is to be omitted from the formal classification that is the problem. Cladists emphasize cladogenesis, omitting anagenesis and grades that are paraphyletic. Evolutionists emphasize evolutionary origins of homogeneous groups, cognizant of the fact that such groups do exist and therefore must be explained, and omit a complete tabulation of cladogenesis.

Classifications are not theories. Several authors of quite diverse views (e.g., D. L. Hull, pers. comm.; Cracraft 1974:79; Ashlock 1979:448; Estabrook 1981:98) have expressed the opinion that the best use of a formal classification is as an index to the analysis from which it is derived and the organisms that it includes. Because formal classifications can only imperfectly express a full phylogenetic analysis, classifications are of less significance to systematics than is progress in analytical methods. It follows that the classification that serves as the best index to the analysis that produced it, and the easiest to use, is the wisest choice.

In 1979 I suggested that homogeneous monophyletic groups are easier to identify, remember, and make generalizations about—good index criteria—than is a collection of strictly holophyletic groups, many of which are far from homogeneous and sometimes more difficult to identify. Many holophyletic groups are of necessity established on weak evidence. I also suggested that because minor errors in cladistic analysis may lead to major changes in cladistic classification, strictly holophyletic classifications are therefore not particularly stable. For these reasons, holophyletic classifications do not serve as well the strictly practical purposes to which classifications are put. In any case, the best repository for what is known about organisms is the phylogenetic analysis, not the formal classification.

Hull wrote: "The desire on both sides to retain the term 'monophyly' for their concept stems from the need to establish continuity in scientific development. Just as continuity matters in individuating lineages in biological evolution, it matters in individuation of a scientific lineage in conceptual evolution" (1979:435). Differing usages of the word monophyly are the result of different fashions, so that the word has, unfortunately, become a symbol of two lines of continuity, and much of the discussion is sociological. Adamant cladists simply chose to accept Hennig's revised concept of monophyly. With others, I prefer the continuity of the traditional concept.

Acknowledgment

I would like to thank Thomas Duncan for his invitation to participate in this symposium and for his highly perceptive editing of the manuscript as submitted.

Literature Cited

Ashlock, P. D. 1971. Monophyly and associated terms. *Syst. Zool.* 20:63–69.
——. 1972. Monophyly again. *Syst. Zool.* 21:430–438.
——. 1979. An evolutionary systematist's view of classification. *Syst. Zool.* 28:441–450.
Bock, W. J. 1977. Foundations and methods of evolutionary classification. In M. K. Hecht, P. C. Goody, and B. M. Hecht, eds., *Major Patterns in Vertebrate Evolution*, pp. 851–895. New York: Plenum Press.
Charig, A. J. 1982. Systematics in biology: A fundamental comparison of some major schools of thought. In K. A. Joysey and A. E. Friday, eds., *Problems of Phylogenetic Reconstruction*, pp. 363–440. London, New York: Academic Press.
Cracraft, J. 1974. Phylogenetic models and classification. *Syst. Zool.* 23:71–90.
Darwin, C. 1959. *Origin of Species: A Variorum Text.* M. Peckham, ed. Philadelphia: University of Pennsylvania Press.

Darwin, F. 1888. *The Life and Letters of Charles Darwin.* Vol. 2. London: John Murray.

Duncan, T. 1980. Cladistics for the practicing taxonomist: An eclectic view. *Syst. Bot.* 5:136–148.

Estabrook, G. F. 1978. Some concepts for the estimation of evolutionary relationships in systematic botany. *Syst. Bot.* 3:146–158.

———. 1981. [Report on] the Willi Hennig memorial symposium, convened by James S. Farris. *Syst. Bot.* 6:95–100.

Farris, J. S. 1974. Formal definitions of paraphyly and polyphyly. *Syst. Zool.* 23:548–554.

Ghiselin, M. T. and L. Jaffe. 1973. Phylogenetic classification in Darwin's monograph on the sub-class Cirripedia. *Syst. Zool.* 22:132–140.

Haeckel, E. 1874. The gastrea-theory, the phylogenetic classification of the animal kingdom, and the homology of the germ-lamellae. E. P. Wright, trans. *Quart. Rev. Microscop. Sci.,* 2d ser., 14:142–165, 223–247.

———. 1894. *Systematische Phylogenie der Protisten und Pflanzen: Erster Theil des Entwurfs einer systematischen Stammesgeschichte.* Berlin: Georg Reimer.

Hennig, W. 1966. *Phylogenetic Systematics.* D. D. Davis and R. Zangerl, trans. Rpt. 1979. Urbana: University of Illinois Press.

Hull, D. 1979. The limits of cladism. *Syst. Zool.* 28:416–440.

———. 1980. Cladism gets sorted out. Review of *Phylogenetic Analysis and Paleontology,* J. Cracraft and N. Eldredge, eds. *Paleobiology* 6:131–136.

Mayr, E. 1974. Cladistic analysis or cladistic classification? *Z. zool. Syst. Evolut.-forsch.* 12:94–128.

Nelson, G. 1971. Paraphyly and polyphyly: Redefinitions. *Syst. Zool.* 20:471–472.

Platnick, N. I. 1976. Are monotypic genera possible? *Syst. Zool.* 25:198–199.

———. 1977. Paraphyletic and polyphyletic groups. *Syst. Zool.* 26:195–200.

Ross, H. H. 1974. *Biological Systematics.* Reading, Mass.: Addison-Wesley.

Simpson, G. G. 1961. *Principles of Animal Taxonomy.* New York: Columbia University Press.

Tuomikoski, R. 1967. Notes on some principles of phylogenetic systematics. *Ann. Entomol. Fenn.* 33:137–147.

Wiley, E. O. 1977. Are monotypic genera paraphyletic: A response to Norman Platnick. *Syst. Zool.* 26:352–355.

———. 1979. An annotated Linnaean hierarchy, with comments on natural taxa and competing systems. *Syst. Zool.* 28:308–337.

PART II

Character Evaluation

INTRODUCTION

Most of the interest in cladistics has centered on the construction of branching sequences. Although this is an obviously important and visually conspicuous part of any phylogeny, the efficacy of any method of cladistics is largely dependent on the adequacy and informational content of the characters and their states used in the analysis. It is very important, there fore, that care and attention be given to the selection of characters, that their states be described effectively, that the homology of these states be assessed, that these states be placed in a logically acceptable character-state network, and that the evolutionary polarities (primitive vs. derived conditions) of these states be determined.

The topic of homology of character states has been viewed as a "quag-mire" for decades (e.g., Davis and Heywood 1963; Inglis 1966), primarily because of the difficulties of circularity of definition, as well as practical problems. Kaplan provides a thorough review of the philosophical under-pinnings of homology and the difficulties of determining homologies in plants, with excellent examples of the utility of developmental anatomical studies.

Several early papers on polarities of character states are known (e.g., Sporne 1948; Maslin 1952; Marx and Raab 1970). Very recently, many more papers have appeared that have revealed a diversity of viewpoints on what kinds of criteria (or arguments) are admissible to determine polarity. De Jong (1980) and Crisci and Stuessy (1980) have reviewed the topic and recog-nized many criteria. Several other papers have followed that have dis-cussed these and other perspectives (Stevens 1980, 1981; Watrous and Wheeler 1981; Wheeler 1981; Arnold 1981). The paper presented here by Stuessy and Crisci summarizes these more recent ideas.

Literature Cited

Arnold, E. N. 1981. Estimating phylogenies at low taxonomic levels. Z. *zool. Syst. Ev-olut.-forsch.* 19:1–35.

Crisci, J. V. and T. F. Stuessy. 1980. Determining primitive character states for phylogenetic reconstruction. *Syst. Bot.* 5:112–135.

Davis, P. H. and V. H. Heywood. 1963. *Principles of Angiosperm Taxonomy.* Princeton, N.J.: Van Nostrand.

De Jong, R. 1980. Some tools for evolutionary and phylogenetic studies. *Z. zool. Syst. Evolut.-forsch.* 18:1–23.

Inglis, W. G. 1966. The observational basis of homology. *Syst. Zool.* 15:219–228.

Marx, H. and G. B. Raab. 1970. Character analysis: An empirical approach applied to advanced snakes. *J. Zool.* (London) 161:525–548.

Maslin, P. P. 1952. Morphological criteria of phylogenetic relationships. *Syst. Zool.* 1:49–70.

Sporne, K. R. 1948. Correlation and classification in dicotyledons. *Proc. Linn. Soc. London* 160:40–47.

Stevens, P. F. 1980. Evolutionary polarity of character states. *Ann. Rev. Ecol. Syst.* 11:333–358.

———. 1981. On ends and means, or how polarity criteria can be assessed. *Syst. Bot.* 6:186–188.

Watrous, L. E. and Q. D. Wheeler. 1981. The out-group comparison method of character analysis. *Syst. Zool.* 30:1–11.

Wheeler, Q. D. 1981. The ins and outs of character analysis: A response to Crisci and Stuessy. *Syst. Bot.* 6:297–306.

4

The Concept of Homology and Its Central Role in the Elucidation of Plant Systematic Relationships

Donald R. Kaplan

Plant developmental morphology, my area of specialization, would seem far removed from cladistics. Nevertheless, since problems of structural homology are central to questions of systematic affinity and phylogenetic relationship, they require insight from a range of specializations. Having been concerned with questions of the developmental bases of the diversity of plant form over the past 20 years, I have had to confront problems of homology in a variety of plant groups to establish that the components I was examining in my developmental studies were indeed structurally comparable.

At a time when cladistic analysis is helping to make phylogenetic reconstruction more rigorous and explicit, it is imperative that the structural analyses which form its foundation be equally critical. Otherwise we risk constructing a phylogenetic "house of cards." In recent years I have become concerned that the rigor of the sophisticated methods of synthesis has begun to outstrip that of the information being synthesized. It is for these reasons that I wish to reinforce the central role of morphological analyses in the characterization of plant evolution.

Critical studies of structural correspondence still provide the crucial first approximation of systematic relationship. But they are equally significant in the coding of character states for cladistic analysis. In the examples that follow I will show how detailed morphological and developmental investigations utilize the concept of homology in reaching their conclusions and how such detailed investigations can provide new insights into problems of organismic affinity.

Because there is some suggestion that determination of the anatomical
(cell to tissue) level of higher plant organization may occur independently
of the organographic level (examples abound of plant taxa with convergent
external morphologies but diagnostic histologies), I shall confine my discussion largely to the organ level of plant organization. But before turning
to specific examples, I would like to deal briefly with the concepts of homology and analogy and the criteria used in the determination of homology.

The Concepts of Homology and Analogy

No concept in biology seems to have engendered so much debate for so
little actual gain as the concept of homology. Over the past few decades
hardly an issue of *Systematic Zoology* has rolled off the press that did not
subject this much maligned and basically misunderstood concept to additional scrutiny and dissection. The disagreement generally concerns not how
homologies are determined but what they mean.

As summarized by Inglis (1966) and by Sneath and Sokal (1973), the debate boils down to whether the concept is defined operationally, i.e., on
the basis of morphological correspondence, or whether it includes the assumption of commonality of ancestry. As Sneath and Sokal (1973) have argued, the weakness of the phylogenetic definitions of Simpson (1961) and
others (Eldredge and Cracraft 1980; Wiley 1981) is that they define homology in terms of the conclusions they wish to reach instead of using an
independently derived concept that is applied secondarily to problems of
phylogeny.

While I realize that most systematists work with taxonomic groups whose
members are assumed to have evolved from a common ancestor, a plant
morphologist has to deal with the problems of homology at the level of first
principles without preconceived assumptions of affinity. This is why it is
necessary to define homologies operationally at first, on the assumption that
when taxonomists combine such morphological data with other sources of
information, they will be able to decide which structural equivalents are
the result of convergence and which represent a homology in a phylogenetic sense. Ideally, taxonomists should be similarly skeptical of past suggestions of relationship within their group of organisms and should likewise start from square one in determining homologies.

A major difficulty in studying the general morphology of higher plants is
that despite some minor, group-specific variants, they all are constructed
on the same general organizational theme or groundplan, regardless of the

specific structural adaptations they may show (Troll 1935–1943). Apparently the evolution of this plan took place so early in the history of vascular plants (Chaloner and Sheerin 1979) and must have been sufficiently basic to the functioning of terrestrial photoautotrophs (Whittaker 1969) that it stabilized early and became fundamental to the plant diversity that radiated from it. For these reasons all vascular land plants exhibit a certain homology of organization, which transcends individual systematic boundaries (Wardlaw 1965). However, within this common groundplan there still are marked structural and functional convergences among individual organ components (e.g., shoot, leaf, and root tendrils). Hence the first goal of any morphological analysis must be to determine which of the fundamental components of the groundplan are elaborated for the specific biological roles of the plant. Because of the occurrence of gross structural convergences between groups that are widely separated systematically (e.g., stem succulents in Cactaceae and Euphorbiaceae), the practicing morphologist must use a homology concept that simply refers to the homology of organization *before* attempting to discriminate the phylogenetic significance of the correspondences.

Some workers have suggested that a different term be used for the general morphological similarity that I have referred to above as a homology of organization (Wardlaw 1965). Hunter (1964), for example, has proposed that the term *paralogy* be used to refer to structural similarity where no phylogenetic conclusions were being drawn. Two structures would then be *paralogous* rather than *homologous,* and the structures themselves would be called *paralogues* rather than *homologues.* In figure 4.1 I have reproduced the schematic diagram depicting the relationship of the concept of paralogy to phylogenetic concepts of homology and analogy originally published by Inglis (1966).

While I agree that the term *paralogy* fills the need for the expression of a structural relationship before the degree of affinity has been established, I am not fond of a needless proliferation of terms. Therefore, I suggest that the term homology should continue to be used to refer to a structural correspondence regardless of its phylogenetic implications and that Lankester's (1870) terms homogeny and homoplasy be used where there is or is not a phylogenetic relationship, respectively. Given the long time since the inception of Lankester's terminology, it is surprising that it has not been used more widely in recent times.

By contrast, definition of the term *analogy* has been far less controversial. Analogy typically refers to the situation in which two entirely different organ components share the same function. For example, root, shoot, and leaf spines are analogous rather than homologous. They all have a spine-

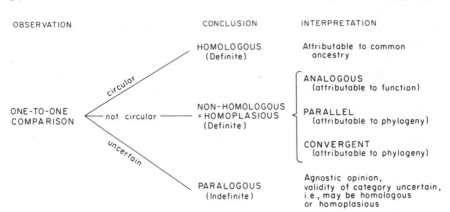

Figure 4.1. Diagram of stages involved in recognition and interpretation of morphological resemblance (reproduced with permission from Inglis 1966).

like structure and/or function but represent different organ components of the plant body that have been equivalently metamorphosed.

Criteria of Homology

What is more important for the practicing systematist than mere matters of definition is the actual criteria that are used to determine whether two seemingly divergent structures are or are not homologous. Despite its importance, this facet of the homology problem rarely receives attention.

In my opinion the clearest delineation of the homology criteria is that given by the German systematist Remane (1952). According to Remane, two structures are considered homologous if they fulfill one or more of the following criteria:

1. Equivalent positions within the general ground plan or organization (Das Lagekriterium)
2. Equivalent special quality (Des Qualitätskriterium)
3. Connection of differing structures by intermediate forms (Das Stetigkeitskriterium)

Both Eckardt (1964) and Hagemann (1975) have given a thorough discussion of the application of Remane's criteria to problems of plant homology. Suffice it to say the criteria should not be viewed as a series of rules to be applied inflexibly to a problematic form. They merely summarize the type of decision making that a morphologist goes through in deciding whether particular organs are or are not structurally comparable.

The positional criterion is self-explanatory and influences many of the distinctive properties of the organ under consideration. For example, in the case of leaves, their dorsiventral transectional symmetry, distinctive growth distribution, and tissue orientation are all undoubtedly products of their lateral insertion on the shoot axis or stem. While there are well-known divergences from fixed positional relationships, especially with regard to lateral branch buds (Sattler 1975), many of these aberrancies are the result of developmental alterations in which a particular organ arises in a conventional position but subsequently is displaced by differential growth, e.g., buds in Cactaceae (Boke 1952) and Vitaceae (Millington 1966).

The criterion of special quality refers to those characteristics of an organ that are unique to it but independent of its position within the total plant body (for example, the distinctive configuration of the vascular core of a root, its possession of a root cap, etc.). Such a criterion allows one to identify an organ component even though it no longer retains an organic connection with the rest of the plant body. This criterion is obviously of utility in dealing with fossil plant material, where the lack of organic connection is a recurrent problem.

The criterion of the connection of two seemingly divergent structural features by intermediates is one of the most useful in determining homologies. In metameric organisms such as plants, the intermediates may be found either along the axis of an individual plant or between different individuals or species of a larger taxonomic unit. When the transitions occur along the length of an individual plant, it is termed "serial homology." A good example of serial homology is the marked change in leaf form exhibited by seedlings of species of *Berberis,* where the first-formed primary leaves are conventionally green and foliar whereas the later-formed are non-green and spiniferous and foliage leaves are borne on axillary short shoots. If one were to see detached pieces of shoot from each of these sectors, one might conclude that they were from distantly related, or unrelated, taxa. But the occurrence of intermediates that have partially foliaceous laminae with spine-tipped apices borne on the same position on the shoot axis links the two leaf forms and establishes their structural relationship.

Structural intermediates can also be found between different species of closely related taxa. For example, in Cactaceae all transitions occur from leafy cacti with large, conspicuously laminate leaves (e.g., *Pereskia*) through those where the lamina region has been reduced and the leaf base enlarged slightly (e.g., *Opuntia*) to those where the laminar region is completely aborted and the leaf base or podarium is expanded and serves as the leaf's major photosynthetic surface (e.g., *Mammilaria*). If it were not for the occurrence of species intermediate in morphology between the more conventional leafy forms and the derived succulent form, interpretation of the *Mammilaria*-type morphology would be much more problematic.

When a change in a character state occurs during the ontogeny of a particular structure, developmental stages can serve as the intermediates that link the two divergent character expressions. For example, in the case of root spines, prior to the sclerification of the root apex, the organ exhibits the typical configuration of a root apical meristem, including a distinctive root cap (McArthur and Steeves 1969). However, as the cells of the apex elongate with the onset of sclerification, the boundary between the root body and cap is obliterated, so that at maturity one cannot tell a root spine from that of any other metamorphosed, spiniferous organ. Hence, in this instance it is the developmental stages that supply the transitions between the more conventional character of a root and the highly altered, seemingly aberrant state of a root spine.

Of course, there are instances where developmental change is congenital, i.e., where an organ or component arises in an anomalous location or exhibits its distinctive properties from inception. Under such circumstances developmental observations obviously are of little aid in establishing structural relationships, and the homologies must be determined from the comparative morphology of mature forms (Kaplan 1971).

Having dealt with the concepts of homology and analogy as well as the criteria used in their establishment in plants, I now turn to specific examples of their application to problems of plant evolution and systematic affinity.

The Contribution of the Concept of Homology to the Elucidation of Evolution in Coniferous Gymnosperms

In my opinion, the investigation that best exemplifies the fundamental role that the solution of homologies plays in phylogenetic research is the classic work by the great Swedish paleobotanist-systematist Rudolf Florin on the evolution of the ovulate cone in conifers (summarized in Florin 1951). Florin's research is a model of comprehensive, analytical-morphological investigation carried out with a meticulous concern for detail coupled with a profound understanding of the principles of plant morphology. For anyone interested in how a study of evolutionary morphology should be executed, I recommend the reading of this classic work.

Prior to Florin's study, one of the most controversial topics in the field of plant morphology was the interpretation of the organography of the ovulate (female or megasporangiate) cone in conifers. The principal argument was whether the female cone was the homologue of a simple strobilus (flower) or an entire flowering branch (inflorescence). The resolution of this question in turn rested on the correct interpretation of the ovuliferous scale complex located in the axil of each bract on the ovulate cone axis. Because of the very simplified and, in some cases, highly modified nature of the ovuliferous scale-ovule complex, there had been at least eight different interpretations of its morphology, ranging from the ridiculous to the sublime.

As a prerequisite to his studies, Florin had the foresight to carry out an extensive survey of epidermal characteristics in the gymnosperms as a whole, with particular attention devoted to the structure of the stomatal complexes and their ontogeny. This epidermal survey not only supplied him with new evidence of systematic alignments within the gymnosperms but also gave him an independent test of affinity for the organisms whose morphology he was assessing, as well as an invaluable test of organic connection between the vegetative and reproductive organs in these groups.

Because Florin knew that the carboniferous Cordaitales was a group of significance in the origin of the conifers, he devoted a great deal of attention to a detailed analysis of both the male and female reproductive shoots of various species of the genus *Cordaianthus*. Not only did Florin embed these fossil remains in hard resin in order to make longisections of the sporophylls and sterile scale leaves; he also cut serial transections through entire reproductive shoots so that he could characterize the phyllotaxis and determine how sterile and sporangia-bearing appendages were related to each other positionally on these axillary shoots. As a result of these remarkably modern phyllotactic analyses he was able to demonstrate unequivocally that sporophylls were not axillary to sterile leaves but were disposed along the same genetic spiral as the sterile appendages. Thus Florin was able to show that both male and female dwarf shoots were homologous with an individual strobilus and that it was the entire reproductive branch bearing these axillary short shoots that was equivalent to an inflorescence.

Using the morphological insights derived from his analysis of the *Cordaianthus* reproductive shoot, Florin proceeded to make a comprehensive study of female cones of fossil conifer taxa extending from the upper Carboniferous and Lower Permian to the Upper Permian and Mesozoic. He was able to demonstrate the following trends in ovulate cone and ovuliferous scale evolution: a reduction in shoot length; a change in shoot transectional symmetry from radial to dorsiventral; a change in phyllotaxis from spiral to decussate; a displacement of remaining sterile scales toward the abaxial side and their adnation into a disclike structure; and an inversion of ovules and their adnation to the flattened sterile component. Thus, it was clear from Florin's studies that the ovuliferous scale is in fact homologous with an axillary reproductive Cordaitalean shoot which in the course of evolution had become markedly reduced in size and flattened and meta-morphosed to such a degree that in its extreme form in contemporary con-ifers it bears little or no resemblance to its fertile shoot homologue in the Cordaitales. Were it not for the existence of intermediates linking the two extremes in morphology, Florin would not have been able to give such an accurate evaluation of the morphological status of the ovuliferous scale and its phylogeny.

What is of prime interest for this discussion is that the success of Florin's effort was based largely on an effective application of the homology crite-ria that I enumerated above. He clarified the basic positional and organi-zational features of the Cordaitalean reproductive shoot and then used that organizational knowledge to interpret each minor variant from the basic groundplan. Moreover, by his effective use of the morphological interme-diates he found in the fossil record, he was able to link structurally two extremes of axillary shoot expression which, if considered without such transitions, would seem to bear no structural relationship at all. The fact that Florin's transitional forms appeared along a time sequence in the fossil record was really "icing on the cake" and not fundamental to his mor-phological conclusions. Had the same intermediates been preserved among contemporary conifers instead of in fossils his conclusions would have been just as valid, because they would still have been based on a fundamental appreciation of morphological principles.

Examples Showing the Need for Sound Morphological Analyses in Systematic Research: Structural Variation in *Acacia*

Having shown how the concept of homology was fundamental to one of the classic studies of plant evolution, I now turn to two recently investi-gated examples from my own research in the genus *Acacia* (Mimosoideae-Leguminosae). The results of these investigations have altered the mor-phological interpretation of *Acacia* species in ways that can have an im-pact on their assessment by a cladistic analysis. In the first case, the prob-lem of the morphological nature of the phyllode in Australian species of the genus presents an example where species that at first glance look dis-tinctly different in form turn out on closer evaluation to be less divergent than they initially seemed. In the second case, that of *Acacia verticillata,* we have just the opposite—what appears superficially to be a relatively uniform type of leaf and shoot morphology turns out under more critical evaluation to be remarkably more complex and divergent. In both cases the new morphological insights acquired call for a renewed assessment of the systematic affinities of the taxa involved.

Turning first to the question of phyllode morphology, it has long been known that in contrast to the bipinnate leaves characteristic of the majority of the species in the genus (Airy Shaw 1966), the 300 or so species of *Acacia* in the section *Phyllodineae* from Australia bear simple, medianly flattened foliar appendages called phyllodes. However, as demonstrated by Cambage (1915–1928), all phyllodineous acacias bear pinnately dissected leaves in their seedlings and then undergo a morphological transition to the simple phyllode borne in later shoot growth. It was the leaf transitions in the seedlings that served as the basis of the deduction of the homologies between pinnate leaves and phyllodes in the genus. Unfortunately, there is a fundamental problem with the nature of the morphological change that has led most investigators to misconstrue the developmental and hence the morphological relationship between phyllodic and pinnate leaves in *Acacia* (Kaplan 1980).

According to the traditional interpretation of phyllode morphology, which was based simply on the comparative morphology of fully mature seedling pinnate leaves, phyllodes, and transition forms between them, the phyllode was derived by the progressive suppression of the distal laminar region, associated with a compensatory elongation and/or widening of the petiole region (fig. 4.2D) to form the leaf's major photosynthetic surface (fig. 4.2L). The bipinnate seedling leaves bearing a single leaflet pair at their apices were considered to represent a transition form in which lamina suppression was incomplete and the petiole only partially expanded (fig. 4.2H). The small, nubbinlike apical pointlet at the tip of the phyllode (a, fig. 4.2L) was assumed to be the remnant of the reduced lamina which remained at the summit of the expanded petiole. This interpretation was based solely on the analysis of mature leaves, assuming a particular mechanism of development. An underlying assumption was that both phyllodes (fig. 4.2L) and bipinnate transition leaves really were pinnate foliage leaves (fig. 4.2D) which went through the same developmental pathway (fig. 4.2A–D) but underwent an arrest in development in the blade region and a compensatory, precocious intercalary expansion of the petiole (fig. 4.2E–H, I–L).

Because pinnate intermediates formed on coppice reversion shoots of *A. melanoxylon* did not conform to the petiolar interpretation (Kaplan 1975), I decided to investigate the comparative development of successive foliar

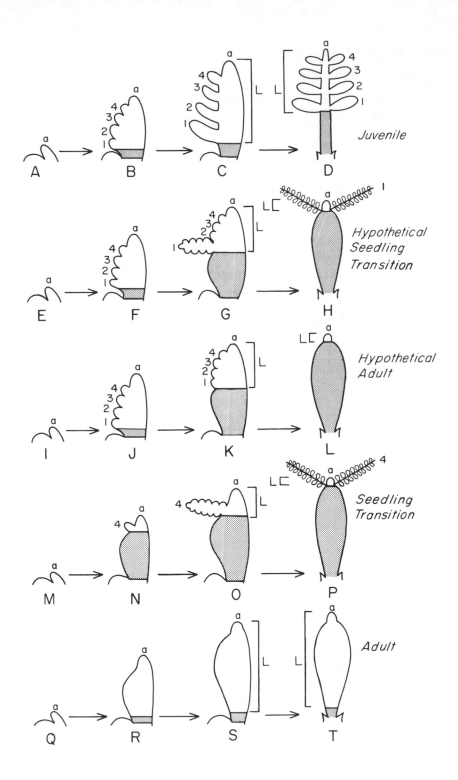

Juvenile

Hypothetical
Seedling
Transition

Hypothetical
Adult

Seedling
Transition

Adult

organs in four phyllodineous species of *Acacia*. The comparative developmental evidence (Kaplan 1980) demonstrated unequivocally that the blade of the phyllode is the positional homologue of the lamina of the pinnate leaf (compare fig. 4.2Q–T with fig. 4.2A–D). Instead of undergoing a growth displacement in which the petiole is extended in place of the superposed lamina (fig. 4.2I–L), the blade sector of the phyllode simply undergoes a unified growth instead of the transverse and longitudinal subdivision into leaflets (compare fig. 4.2Q–T with 4.2A–D and with fig. 4.3A, B and G, H). The sharp contrast in mature form between pinnate and phyllodic organs (fig. 4.3K, L) is simply the result of the mature phyllode retaining the vertical planation that was characteristic of its early adaxial-abaxial plane of growth, whereas the pinnate leaves assume a more conventional dorsiventral planation as a result of pulvinar rotation of their leaflets (compare, in fig. 4.3, C with D, E with F, and K with L).

Furthermore, by actually comparing the developmental properties of a heteroblastic leaf series in a phyllodineous species (e.g., *A. longifolia*) with that in a species with a prolonged pinnate seedling phase (e.g., *A. melanoxylon*), I was able to demonstrate indirectly that blade length actually increased rather than decreased in successive leaves in the phyllodineous taxa in the same way that it does in the more conventional bipinnate species. It is only because the early change in blade form (from pinnate to phyllodic form) eliminates the boundary between blade and petiole in phyllodineous species that previous workers misinterpreted the changes in proportion that actually take place.

At the very least, the results of these investigations emphasize that simply looking at transitional forms between character states (homology criterion 3) may not be sufficient to resolve the homology between two structures when some feature of the change actually obscures the fundamental nature of the structural relationship. It then becomes necessary to look more deeply in order to effectively resolve the real relationship.

Figure 4.2. Diagrammatic representation of the developmental relationships among a pinnate leaf, transition form, and phyllode in *Acacia*. A–D, developmental stages of a once-pinnate juvenile leaf; E–H, hypothetical developmental stages of a bipinnate transition leaf with a petiolar origin; I–L, hypothetical developmental stages of a phyllode with a petiolar derivation; M–P, actual developmental stages of a seedling transition; Q–T, actual developmental stages of a phyllode. Shaded areas are petiolar derivatives; unshaded areas are lamina derivatives and stipular base. The numbers 1–4 give the sequence of leaflet initiation; the letter *a* indicates the apical pointlet and *L* the lamina. Since it is not possible to indicate the vertically extended nature of the phyllode in an adaxial view, H, L, P, and T show the phyllodic blade region expanded from an adaxial view even though the plane of enlargement is actually at right angles to the plane shown (reproduced with permission from Kaplan 1980).

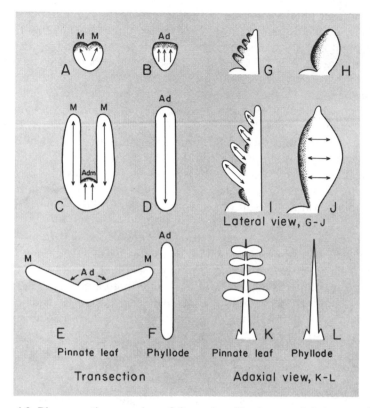

Figure 4.3. Diagrammatic comparison of the modes of lamina growth between a phyllode and pinnate leaf in *Acacia*. A–F, transectional views. G–J, median radial longitudinal views. K, L, adaxial views of mature pinnate leaf and phyllode, respectively. Stippled areas are regions of meristematic activity. Arrows give the principal direction of growth. *Ad*, adaxial; *Adm*, adaxial meristem; *M*, marginal meristem (reproduced with permission from Kaplan 1980).

But beyond these resolutions of homology and phyllode developmental divergence, the results of this morphological investigation can have broader systematic implications. By demonstrating that the phyllodineous species undergo the same general ontogenetic changes in size and proportion of blade to petiole, it makes the phyllodineous taxa seem far less divergent than the previous morphological interpretations have suggested. Moreover, because it could be shown that the actual developmental differences between phyllodic and pinnate laminae are simply a matter of dissected versus unified growth, it means that the contrast between these two organ types is not nearly as great as suggested by mature morphology. Given that these new insights suggest that there is less difference between the two sections of the genus than previously suggested, such new morphological information is bound to have an impact on subsequent systematic treatments.

The second problematic taxon in this genus, *Acacia verticillata*, presents the opposite problem from that of the phyllodineous species just summarized. Whereas in the latter a marked difference in leaf form turned out not to be as fundamental as initially thought, what appears at first to be a uniformity of leaf (phyllode) form in *A. verticillata* turns out to be markedly heterophyllodic, with appendages that are not homologous with one another.

Like other phyllodineous acacias, *A. verticillata* exhibits a typical heteroblastic seedling development, with a series of pinnate leaves and transitions initiated prior to the differentiation of the needlelike phyllodes that are borne for the remainder of shoot growth (fig. 4.4). Two distinct types of phyllodes are actually produced at regular intervals. One type bears stipules, has an extrafloral nectary on its adaxial surface, and has a shoot bud in its axil. The other is exstipulate and lacks both nectary and axillary bud. The exstipulate appendage is only slightly smaller than the stipulate but outnumbers the latter by a ratio of 8:1. Stipulate phyllodes can be distinguished from the exstipulate organs macroscopically because of the presence of a conspicuously expanded shoot in their axils (figs. 4.4 and 4.8).

It should be mentioned that the heterophyllodic nature of *A. verticillata* has been known for some time and was studied by both Hofmeister (1868) and Goebel (1905, 1928). Hofmeister considered the exstipulate organs to be stipular appendages which came to resemble the true leaf homologues (the stipulate phyllodes) in the course of development, whereas Goebel (1928) considered all the appendages to be leaf homologues which differed simply in their early rates of growth. Hara and Kaplan (1980) decided to reinvestigate the details of its developmental morphology because of the sharp conflict in interpretation between these two noted workers.

Our observations (Hara and Kaplan 1980) indicate that the primordia of the two phyllode types not only are distinctly different in size and shape when they arise but are initiated from distinctly different regions of the shoot axis and have a different type of vascular supply. Whereas the stipulate phyllodes are initiated as broadly flattened primordia in a 2:3 phyllotactic system at a divergence angle of 137.5° (Kaplan 1980), the exstipulate organs show no relation to this classic helical pattern. The exstipulate appendages arise as much smaller, globose outgrowths in a transverse sequence beginning at the margins of the stipules of the stipulate phyllode and then proceeding circumferentially from a meristem-encircling, collarlike extension of the leaf base (figs. 4.5 and 4.6). As the stipulate leaf enlarges lat-

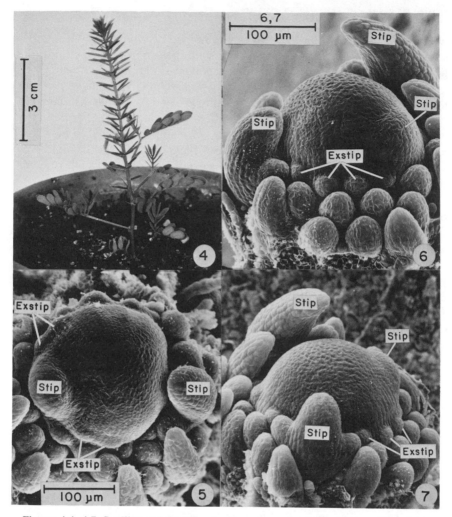

Figures 4.4–4.7. Seedling and scanning electron microscope views of shoot apices of *Acacia verticillata.* **4.4. Seedling showing the pinnately dissected leaves characteristic of early seedling growth and needlelike phyllodes characteristic of later shoot growth (note that except for the position of axillary shoots, it is virtually impossible to distinguish stipulate from exstipulate phyllodes in this view). 4.5. Apical view of meristem showing the most recently initiated stipulate phyllode and the earliest inception of exstipulate phyllodes emerging as small protuberances from a circumferential leaf base expansion around the periphery of the shoot apex (note the marked differences in size and shape of the exstipulate phyllode organs from those of the stipulate appendages, as well as the differences in relative position of insertion on the meristem proper). 4.6. Semilateral view of shoot apical meristem showing young stipulate phyllodes and exstipulate appendages emerging from the meristem periphery. 4.7. Semilateral view of the shoot tip region showing recently initiated stipulate phyllodes with their clearly developed lateral stipules and their contrast with the much smaller protuberances which are the exstipulate organs.** *Exstip,* **exstipulate appendages;** *Stip,* **stipulate appendages.**

erally, the tiny exstipulate organs associated with it continue to be initiated around the circumference of the stem until they encroach on the next oldest stipulate organ (figs. 4.6 and 4.7). Since the exstipulate appendages arise after the associated stipulate phyllode, they are always initiated some distance down the flanks of the shoot axis, in contrast to the more apical-lateral site of stipulate phyllode inception (figs. 4.6 and 4.7). Interestingly, while the primordia of both stipulate and exstipulate organs look vastly different when they arise (figs. 4.5–4.7), they subsequently come to resemble one another to the extent that in a transection of the terminal bud they can barely be distinguished from one another (figs. 4.8, 4.24, and 4.25).

The stages of this remarkable developmental convergence are best appreciated from a comparison of median to near-median longisections through successively older (longer) stages of stipulate, exstipulate, and stipular organs in figs. 4.9–4.23. In all the sections the stipulate phyllode can be recognized by the presence of a meristem in its axil from an early stage on (figs. 4.11, 4.16–4.18), whereas the exstipulate organ never shows a sign of an axillary bud (figs. 4.13, 4.14, 4.19–4.21). The scanning electron micrographs show not only that the exstipulate appendage is distinctly smaller when it arises but also that it is consistently narrower and less hyponastically arched than its stipulate counterpart at equivalent stages of development (compare figs. 4.12–4.14 with figs. 4.10 and 4.11). In its earliest stages the exstipulate organ is histologically more similar to the stipules of the stipulate phyllode than it is to its leaf axis (compare figs. 4.14 and 4.19 with figs. 4.15 and 4.22).

By the time it reaches a length of about 300 μm, the exstipulate organ has undergone an equivalent amount of radial growth in its basal blade region (fig. 4.20). Despite having a spiniferous apex of greater proportion to its total length, at progressively older stages the development of the exstipulate blade comes to resemble that of the stipulate leaf more than that of its stipules (compare figs. 4.20 and 4.21 with figs. 4.17 and 4.18 and figs. 4.22 and 4.23). Hence, except for the presence of the pair of stipules and an axillary bud, it is virtually impossible to distinguish stipulate from exstipulate organs in transection in later stages of development (compare fig. 4.24 with fig. 4.25). However, despite this ontogenetic convergence the two appendages can be distinguished by their respective vascular supplies. The stipulate organs are innervated from a three-traced unilacunar node, whereas the exstipulate is supplied by a single trace without any gap.

Our survey of the morphological relationships between other leaf types and exstipulate appendages, including juvenile pinnate leaves and bud scales, has demonstrated that each true leaf or "phyllome" homologue has a number of exstipulate outgrowths associated with it. And, as in the association

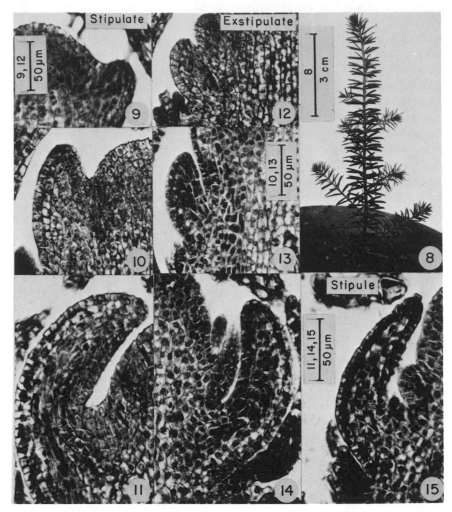

Figures 4.8–4.15. Shoot morphology in *Acacia verticillata*, along with a developmental comparison between stipulate and exstipulate appendages and stipules at nearly comparable developmental stages. 4.8. Seedling showing extensive branch development; sites of axillary branches demarcate the positions of stipulate phyllodes along the length of the main axis (from this comparison it can be seen that the exstipulate organs greatly outnumber the stipulate ones). 4.9–4.11. Median longisections of successively older stipulate leaf primordia of the following lengths: 4.9, 20 μm; 4.10, 40 μm; 4.11, 100 μm. 4.12–4.14. Median longisections of successively older exstipulate appendages of the following lengths: 4.12, 20 μm; 4.13, 45 μm; 4.14, 100 μm. 4.15, near-median longisection of a stipule (from a stipulate leaf) 95 μm in length.

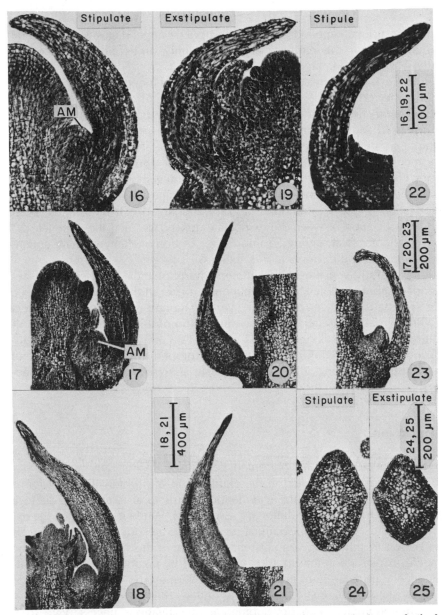

Figures 4.16–4.25. Developmental comparisons of older stipulate, exstipulate, and stipule appendages in *Acacia verticillata*. 4.16–4.18. Median longisections of successively older stipulate leaves of the following lengths: 4.16, 250 μm; 4.17, 500 μm; 4.18, 900 μm. 4.19–21. Median longisections of successively older exstipulate appendages of the following lengths: 4.19, 250 μm; 4.20, 500 μm; 4.21, 750 μm. 4.22, 4.23. Near-median longisections of stipules 260 μm and 400 μm in length, respectively. 4.24, 4.25. Transections of stipulate and exstipulate appendages 1560 μm and 1608 μm in length, respectively. *AM*, axillary meristem.

with the stipulate appendage, these exstipulate appendages arise as lateral outgrowths from the leaf base region of juvenile and scale leaves after the latter have been initiated as distinct organs.

On the basis of these observations, we concluded in agreement with Hofmeister (1868) that the so-called "exstipulate phyllodes" are really supernumerary leaf base outgrowths more closely homologous with metamorphosed stipules than with the foliage leaf axis itself. They are initiated from a circumferential expansion of the leaf base *after* the upper leaf zone (primordial lamina-petiole region sensu Eichler 1861) has been initiated. They come to resemble true leaves because they show a progressive morphogenetic and histogenetic convergence with stipulate leaves and their stem elongation is both "intranodal" and internodal, so that the original positional association between the exstipulate appendages and the "true leaves" of the shoot is lost, giving an impression of homogeneity of leafy appendages borne on the shoots of this species.

Obviously, given the vastly different morphological statuses of the appendages that superficially resemble one another in this species, the search for systematic alliances between *A. verticillata* and other species of *Acacia* could be markedly affected by the affirmation of this type of shoot organization. Hence, this is one more example of a case where rigorous specification of organ homologies can have a profound impact on the nature of a cladistic study, beginning with the basic analysis and coding of characters.

Conclusion

The selected examples presented here illustrate not only how the homology criteria are applied in the clarification of problematic plant forms but also how important the elucidation of structural relationships is to a proper systematic and evolutionary assessment. While it is true that the examples from *Acacia*, especially *A. verticillata*, are quite unusual, there is no reason to think they are not representative of the kinds of problematic species construction that may go undetected by taxonomists in their treatments of particular plant groups. The peculiarities and complexities exhibited by such problematic taxa merely underscore the kind of comprehensive morphological and developmental investigations necessary to truly understand the organization of these distinctive plants.

To the systematist—evolutionary biologist interested mainly in the classification of a group and/or the reconstruction of its phylogeny, these detailed structural analyses may seem like an enormous effort for a very small dividend. Hence, there is little doubt that they will remain in the province of the plant morphology specialist. Nevertheless, as evolutionary biologists have continued to deepen their interest in problems of the origin of species and adaptations, they have become increasingly interested in the developmental bases of the origin of species (Gould 1977). Thus, in the same way that genetic, cytological, and biochemical approaches have become routine tools for the evolutionary biologist of today, it is possible that developmental morphological studies of the kind outlined here will ultimately be incorporated as well. If this comes to pass, I feel confident that the systematists of tomorrow will develop a more profound understanding of the characters they are attempting to assess in their taxonomic treatments.

Acknowledgments

I wish to express my sincere appreciation to Drs. Lewis Feldman and Molly Whalen and to Ms. Eleanor Crump for their critical reading of the manuscript for this paper and their helpful comments for its improvement.

Literature Cited

Airy Shaw, H. K. 1966. *A Dictionary of the Flowering Plants and Ferns.* 7th ed. Cambridge: Cambridge University Press.

Boke, N. H. 1952. Leaf and areole development in *Coryphantha. Amer. J. Bot.* 39:134–145.

Cambage, R. H. 1915–1928. *Acacia* seedlings I to XIII. *Proc. Roy. Soc. N. S. W.,* vols. 49, 60, and 62.

Chaloner, W. G. and A. Sheerin. 1979. Devonian macrofloras: The Devonian system. *Paleontology* 23:145–161.

Eckardt, T. 1964. Das Homologieproblem und Fälle strittiger Homologien. *Phytomorphology* 14:79–92.

Eichler, A. W. 1861. Zur Entwicklungsgeschichte des Blattes mit besonderer Berücksichtigung der Nebenblatt Bildungen. Inaugural diss. University of Marburg, W. Germany.

Eldredge, N. and J. Cracraft. 1980. *Phylogenetic Patterns and the Evolutionary Process.* New York: Columbia University Press.

Florin, R. 1951. Evolution in Cordaites and Conifers. *Acta Horti Bergiani* 15:285–388.

Goebel, K. von. 1905. *Organography of Plants II: Special Organography.* I. B. Balfour, trans. Oxford: Oxford University Press.

——. 1928. *Organographie der Pflanzen: 1, Allgemeine Organographie.* Jena: Gustav Fischer Verlag.

Gould, S. J. 1977. *Ontogeny and Phylogeny*. Cambridge: Harvard University Press, Belknap Press.

Hagemann, W. J. 1975. Eine mögliche Strategie der vergleichenden Morphologie zur phylogenetischen Rekonstruktion. *Bot. Jahrb. Syst.* 96:107–124.

Hara, N. and D. R. Kaplan. 1980. The problem of phyllode dimorphism and homology in *Acacia verticillata*. *Bot. Soc. Amer. Misc. Ser Publ.* 158:48.

Hofmeister, W. 1868. *Allgemeine Morphologie der Gewächse*. Leipzig: Wilhelm Engelmann.

Hunter, I. J. 1964. Paralogy, a concept complementary to homology and analogy. *Nature* 204:604.

Inglis, W. G. 1966. The observational basis of homology. *Syst. Zool.* 15:219–228.

Kaplan, D. R. 1971. On the value of comparative development in phylogenetic studies—a rejoinder. *Phytomorphology* 21:134–140.

———. 1975. Comparative developmental evaluation of the morphology of unifacial leaves in the monocotyledons. *Bot. Jahrb. Syst.* 95:1–105.

———. 1980. Heteroblastic leaf development in *Acacia:* Morphological and morphogenetic implications. *Cellule* 73:135–203.

Lankester, E. R. 1870. On the use of the term homology in modern zoology and the distinction between homogenetic and homoplastic agreements. *Ann. Mag. Nat. Hist.*, 4th ser., 6:34–43.

McArthur, I. C. S. and T. A. Steeves. 1969. On the occurrence of root thorns on a Central American palm. *Can. J. Bot.* 47:1377–1382.

Millington, W. F. 1966. The tendril of *Parthenocissus inserta:* Determination and development. *Amer. J. Bot.* 53:74–81.

Remane, A. 1952. *Die Grundlagen des natürlichen Systems, der vergleichenden Anatomie und der Phylogenetik*. Leipzig: Akademische Verlagsgesellschaft.

Sattler, R. 1975. Organsverschiebungen und Heterotopien bei Blütenpflanzen. *Bot. Jahrb. Syst.* 95:256–266.

Simpson, G. G. 1961. *Principles of Animal Taxonomy*. New York: Columbia University Press.

Sneath, P. H. A. and R. R. Sokal. 1973. *Numerical Taxonomy*. San Francisco: W. H. Freeman.

Troll, W. 1935–1943. *Vergleichende Morphologie der höheren Pflanzen*. Vol. 1, *Vegetationsorgane*, parts 1, 2, and 3. Berlin: Gebrüder Borntraeger.

Wardlaw, C. W. 1965. *Organization and Evolution in Plants*. London: Longmans and Green.

Whittaker, R. H. 1969. New concepts of kingdoms of organisms. *Science* 163:150–159.

Wiley, E. O. 1981. *Phylogenetics: The Theory and Practice of Phylogenetic Systematics*. New York: Wiley.

5

Problems in the Determination of Evolutionary Directionality of Character-State Change for Phylogenetic Reconstruction

Tod F. Stuessy and Jorge V. Crisci

In systematics, considerable interest has been directed in recent years toward the concepts and methods of phylogenetic reconstruction. Since Darwin (1859), workers have attempted to reconstruct phylogenies of various plant and animal groups. It has been only within the past decade, however, that interest has focused on more explicit and objective ways to construct these branching diagrams. The concepts and techniques for reconstructing this branching aspect of phylogeny have come to be known as cladistics (Camin and Sokal 1965; Mayr 1969).

For most cladistic approaches, decisions on evolutionary directionality of character states must be made. In some methods only branching networks are produced (e.g., Kruskal 1956; Prim 1957; Nelson and van Horn 1975; Felsenstein, this volume), which can be rooted at the end of the analysis by designating one of the taxa on the network as most primitive. This avoids explicit determinations of which character states are primitive and which are derived (at least initially). Most other methods, however, require determination of evolutionary directionality of character states before the analysis can be begun, and this involves applying criteria for primitiveness.

Recently, several workers have attempted to summarize the current criteria for primitiveness (Crisci and Stuessy 1980; De Jong 1980; Stevens 1980; Arnold 1981). In our own recent paper (Crisci and Stuessy 1980), we recognize nine criteria: six that require no prior acceptance of primitive conditions (first-level) and three that do require it (second-level). These are shown here in diagrammatic form in figures 5.1–5.9. The important point of these studies is that the focus is now on particular criteria by which decisions on evolutionary directionality of character states can be made. Just saying a state is primitive is no longer adequate; some criterion or group of criteria must be clearly applied in each case.

Many additional discussions have recently ensued over specific criteria. Two of these criteria are very closely related conceptually: in-group and out-group analyses. Both have been used extensively in cladistic studies, but different points of view still exist on how they should be defined and

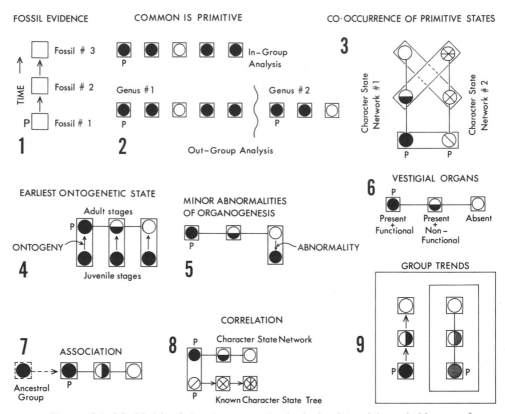

Figures 5.1–5.9. Models of the nine accepted criteria for determining primitiveness of character states (from Crisci and Stuessy 1980). First-level criteria are shown in 5.1–5.6, second-level criteria in 5.7–5.9. In all cases the quadrangles refer to taxa and the circles to characters. Symbols within the circles refer to different character states. Lines connecting character states and taxa indicate phenoclines, with arrows showing directionality (polarity). "P" indicates the primitive character state (or states, in 5.1) of a phenocline.

in what context they should be employed (e.g., Crisci and Stuessy 1980; Stevens 1980, 1981; Watrous and Wheeler 1981; Wheeler 1981). Furthermore, recent comments have also been offered on criteria for fossil evidence (Patterson 1981), vestigial structures (Scadding 1981), and ontogenetic data (Rieppel 1979). To help understand these recent perspectives on polarity of character states, in this paper we offer a review of the different recent ideas on criteria for evolutionary directionality and comments on problems with some of these approaches.

Recent Ideas on Evolutionary Directionality of Character States

Since our paper (Crisci and Stuessy 1980) was published, several additional papers have appeared on the criteria of evolutionary directionality, either independently or in response to our own work. It seems useful, therefore, to review these new ideas as a prelude to offering our perspective on a reasonable (or eclectic) approach to determining evolutionary polarity.

Three new papers have appeared that have reviewed all criteria for determining evolutionary directionality of character states: De Jong (1980); Stevens (1980); and Arnold (1981). De Jong's paper appeared while ours was in press and offers a complex view of the use of different types of directional arguments for evolutionary change. He recognizes several levels of confidence in arguments for determining polarity. The first includes 7 criteria that are "consistent with the evolutionary theory and can give a clue to the direction of evolutionary change" (p. 20): (1) outgroup comparison; (2) vicariance (all states of newly vicarying species will be viewed as apomorphic; this seems to us a variation of the criterion of geographic restriction, which De Jong later rejects in the same paper); (3) geological character precedence (fossil evidence); (4) specialization; (5) predation; (6) emancipation; and (7) retraction and protection. (Criteria 4–7 are valid ecological considerations that, to us, appear to be forms of common-is-primitive in-group analysis.) He also accepts another category of arguments that are "consistent with the evolutionary theory but do not give themselves a clue to the direction of change" (p. 20; similar to our second-level criteria): (1) economization; (2) correlation of transformation series; (3) parallelism in sister groups (our Group Trends); and (4) chorological progression. He gives another grouping of criteria which he believes "are not supported by the evolutionary theory and should not be used" (p. 21): (1) ingroup comparisons; (2) intersections of phenoclines; (3) two identical extremes of two phenoclines; (4) precurrence (= interlocking of phenoclines); and (5) geographic restriction (meaning that endemic taxa are derived). He also treats criteria relying on ontogenetic data as weak. A summary of De Jong's ideas, therefore, indicates that he recognizes as useful 11 different criteria and assigns differing importance to them; he places most confidence in the first 7, however.

Stevens' paper (1980) is very similar to ours. Where we recognize first- and second-level criteria, he uses "independent" and "dependent" criteria, respectively. Our "Fossil Evidence" is his "Paleontology," our "Common is Primitive" his "Character-State Distributions," our "Group Trends" his "Character Sequences and Trends," and so on. The noteworthy feature of the two papers is their similarity, although he (Stevens 1981) has made a strong effort to point out how they differ. His general perspective was similar to ours: "Although none of the dependent criteria should be used alone to assign evolutionary polarity to character states, the question of the circumstances under which other criteria may be used when out-group analysis is impossible is not easy to answer. As in other aspects of systematics it is not that one method is right and the others wrong, but that all are more or less effective; we are dealing with probabilities" (p. 352). One difference (virtually the only one) between our approaches, however, was his rejection of in-group analysis as a valid criterion and placing more emphasis on out-group comparison than on the other first-level criteria.

Arnold (1981) discusses cladistic analysis from many aspects and treats polarity decisions in detail. He stresses the importance of out-group analysis but recognizes the several different kinds of out-group situations that one may encounter (similar to our perspective in Crisci and Stuessy 1980). His summary is: "Of the various supposed polarity indicators, the ones that may provide correct information frequently enough to be of some use are: distribution in outgroup, frequency within studied groups, non-coinciding minority states, correlation with states of other characters, ontogenetic clues, functional clues and fossil evidence" (p. 15).

Developmental criteria have been criticized specifically by some workers. Rieppel (1979) stated, "In the present contribution I will argue, that the comparison of ontogenetic stages follows the criteria of outgroup comparison, and that features of early ontogenetic stages cannot be accepted as indicators of primitive character-states of an adult organism on an *a priori* basis" (p. 57). Scadding (1981) believes that the presence of vestigial structures is not a valid criterion for the following reasons: "Since it is not possible to unambiguously identify useless structures [because you can never know that something does *not* have a function], and since the structure of the argument used is not scientifically valid, I conclude that 'vestigial organs' provide no special evidence for the theory of evolution" (p. 176).

Fossil evidence, to us the strongest evidence of evolutionary directionality (Crisci and Stuessy 1980), has been continually criticized, most recently by Patterson (1981). He stresses that fossils cannot by themselves be used to establish polarities of character states and adds that "it follows that the widespread belief that fossils are the only, or best, means of determining evolutionary relationships is a myth" (p. 218). He acknowledges, however, that "when contradictions [with hypotheses of relationships from recent forms] do come from fossils, they may reverse decisions on homology or on polarity" (p. 219). Fossils, therefore, are judged as secondary aids to polarity determination, rather than primary ones.

Several critiques of our paper have appeared which contain new perspectives. Stevens (1981) criticized our paper from the puzzling perspective that "the approach that I advocated in determining which character states are derived was almost the reverse of that of Crisci and Stuessy" (p. 188). He continued, "This evaluation [of individual criteria] led to the more or less decisive rejection of all criteria except for out-group analysis on the grounds that they were impracticable, depended on other criteria, or led to blatant circularity and so biased the results." Although these comments are not borne out by examination of the papers involved, the critique does represent a new perspective, that the only valid criterion is out-group analysis, although he admits "its use in botany may not be easy."

Watrous and Wheeler (1981) erroneously suggest that our concept of outgroup is not correct and then contrast in-group and out-group analyses, with a rejection of the former and strong acceptance of the latter. The follow-up critique by Wheeler (1981) goes even further and concludes that "all of the criteria discussed by Crisci and Stuessy are either variations of a single concept (out-group comparison) or are not valid, independent criteria for the polarization of character states" (p. 305).

Problems with Criteria of Evolutionary Directionality

A point from our 1980 paper that needs to be reemphasized is that *no criterion* of evolutionary directionality of character states will be absolute. All have conceptual and/or methodological difficulties of varying types. Because we will never know a complete phylogeny (the closest we might come is in a group with an extensive fossil record), it is impossible to know if a criterion is providing accurate information about the true phylogeny or not. It *is* true, of course, that the oldest character state is the most primitive, but we will never know for certain if we have the oldest state represented in the fossil record (although we might wish to assume this in certain instances). Our general perspective, therefore, was, and still is, as follows: "Viewed within this [philosophical] context, therefore, any general criterion for selecting primitive states of a character will at best be only methodologically immanent. That is, it will not be absolute or universal but simply a statement of high explanatory value in which exceptions should be expected and tolerated. It is impossible to do better than this" (Crisci and Stuessy 1980:116).

Despite this, some workers have attempted to demonstrate that all criteria either are forms of out-group comparison (which alone stands unassailable; see, e.g., Watrous and Wheeler 1981; Stevens 1981; Wheeler 1981) or involve severe difficulties. To stress this point, criticisms have been leveled against all other criteria (already discussed), with a special negative tone toward in-group analysis, which has strong conceptual similarities to out-group comparison. We do not believe it useful to address all the specific criticisms raised by these workers; we refer the reader to the literature cited here, plus our earlier paper, where many of these same arguments are addressed.

It is clear that much difference of opinion exists, with some workers preferring narrow approaches and others a broader set of criteria. We stress that the broadest type of evolutionary reasoning should be used before final decisions are made about which criteria are most applicable to a particular group (e.g., some groups have no close relatives, others have no fossils or ontogenetic data, some have no clear phenetic framework within which to begin working, still others are known to have numerous parallelisms). Because so much concern has been raised about in-group versus out-group analyses, however, we believe it most useful to comment in this paper on problems with these two criteria.

In-Group Analysis

The difficulty with defining the in-group criterion is that the concept has been used in many ways in the past. The three main concepts are the strict numerical common-is-primitive idea, i.e., that the primitive character state must occur in a majority of the taxa of the study group; that the primitive character state occurs in a majority of phenetic (or phyletic) lines; and that primitive states are those that seem ecologically generalized. In our opinion, the second definition is the most useful criterion in most cases.

The earliest-used concept of in-group analysis was that the widespread character state in a study group is the most primitive (Frost 1930). From this idea came Danser's groundplan concept (1950), in which the radiation of several taxa from a common ancestor allows the detection of their ancestry by examination of the distributions of character states. In each evolutionary line a different character-state change occurs (perhaps reflecting adaptive radiation into different environments). Note that this does *not* assume dichotomous speciation. This concept was later incorporated as part of the reasoning of the Wagner groundplan-divergence method (Wagner 1961). The simplest (most parsimonious) way of explaining the distribution of character states within a study group of this type, therefore, is to regard the common condition as having been derived from the common ancestor.

Another in-group concept is that a character state is primitive if it occurs in a majority of phenetic (or phyletic) lines within the study group. Here, occurrence within every taxon is not as important as the evolutionary line in which the states are found. For example, in *Melampodium* (Compositae), sect. *Melampodium* contains 20 of the 37 recognized species of the genus (Stuessy 1972), and on that basis all states in that section might be regarded as primitive if simple majority occurrence is followed. A more useful concept, however, is that the state is distributed in a majority of evolutionary lines, in this case four of the six sections (this criterion was actually used in the cladistics of *Melampodium* in Stuessy 1979). The evolutionary reasoning for this concept is that a character state occurring in many of the evolutionary lines of the study group has come from the common ancestor and reflects the primitive condition. This is the same kind of reasoning used for justification of the out-group concept.

The in-group concept, based on ecological generalization (involving distributions, adaptations of structure, etc.), is justified evolutionarily by the idea that a newly evolved taxon initially fills a new ecological zone or niche. Entrance into such a new zone usually is accompanied by structural adaptations that are viewed as specializations, and hence as derived. The common structures in the rest of the taxa are seen as retaining the ancestral (or primitive) condition. Whole groups can evolve together, however, and this perspective must therefore be used cautiously.

Two important points about in-group analysis need to be mentioned. First, in the construction of cladograms using compatibility methods, Estabrook (1977) found that the generated trees were more compatible if common is primitive (simple majority occurrence) than if the reverse is true. In other words, this concept is not likely to be strongly at odds with the way evolution has actually proceeded. Second, in our own experience, polarities derived from in-group analysis usually correlate with those from out-group comparison, which again suggests that the former *are* giving useful information.

The strength of the in-group concept depends to some extent on the ev-

olutionary age of the study group and the kinds of evolution that have occurred within it (the same can be said, however, for other criteria). Older
groups and those that have undergone numerous speciation events, i.e., larger
and diverse taxa, are less likely to still preserve the ancestral condition as
a majority occurrence. With smaller and younger taxa the common patterns should more accurately reflect the ancestral condition. Direct evidence of age of a group is often not available, but morphological, geographical, and ecological diversity can provide clues.

The in-group criterion, like any other criterion, can be misleading in some
cases (Crisci and Stuessy 1980). For example, in the angiosperms, the features of the primitive Magnoliidae are not found in all the other subclasses
of the flowering plants. Some basal groups of the Rosidae and Hamameli-
dae do reflect primitive features (see, e.g., Cronquist 1968, 1981), but these
character states are not pervasive throughout these other subclasses. Without a strong microfossil record (see, e.g., Doyle 1978), we might have difficulty rooting the morphological network, especially because the out-group
is not yet clear (unless it is taken in the broad and therefore minimally informative sense of all the gymnosperms, which are themselves probably
polyphyletic; Delevoryas 1979). Numerous other examples exist, but they
usually occur in groups of considerable evolutionary age and diversity or
in those with clear pectinate (or pinnate) phylogenies (Arnold 1981).

The determination of the occurrence and correlation of character states
within a study group is very important for understanding its evolution. Even
if out-group concepts are employed (as they should be if the out-group can
be clearly identified), it is essential to consider the distribution of character
states within the study group. Ordinarily the two correlate positively. If they
do not, then correlation with ecological and distributional data may give
further clues to the evolution of the character states and taxa in what is
clearly a more complex (but evolutionarily more interesting) situation.

Out-Group Analysis

The out-group criterion also has been used in several different ways. These
assume a character state (one of at least two occurring) within the study
group (or in-group) to be primitive if it is also found in a majority of taxa
of the out-group (= sister group sensu Hennig 1966, i.e., the closest relative at the same rank); in a majority of phyletic (or phenetic) lines of the
out-group; or throughout *all* members of the out-group. The last and most
restrictive definition is the one most recently used by New York Cladists or
Neocladists (so labeled by Van Valen [1978] and Cartmill [1981], respec-

tively; see Watrous and Wheeler 1981). We believe that the second con-cept is the most useful in practice.

The evolutionary reasoning behind the three concepts of the out-group criterion is that a character state is primitive if it occurs in several phenetic or phyletic lines of the study group and related taxa that all come from a common ancestor. Barring reversals or parallelisms, the majority or abso-lute occurrence of a state in different evolutionary lines is presumed to re-flect descent from the common ancestor.

The out-group criterion is a very useful check on the results of the in-group analysis, because it represents an independent assessment of primi-tiveness. Often the results of the in-group and out-group analyses correlate (as they did in 16 of 18 characters when these criteria were applied to the polarity of character states in *Melampodium;* Stuessy 1979). If they do not, explanations can be sought, which should lead to a better understanding of the evolution within the study group. The evolutionary reasoning for the out-group criterion is good because it relies on the generally accepted idea of descent of primitive features from a common ancestor.

Some of the critics of the in-group criterion (e.g., Watrous and Wheeler 1981; Stevens 1981; Wheeler 1981) seem distressed by the idea that out-group and in-group may be closely related conceptually. By strongly op-posing one of the concepts, therefore, one of necessity is also criticizing in some measure the other. Watrous and Wheeler state that "the topic of character polarity cannot be effectively dealt with in the absence of some reference to the commonality principle, which has received and continues to receive support in systematics" (1981:10). Platnick comments, "But does either kind of test [ontogeny or out-group comparison] actually demon-strate one character state to be more primitive in real evolutionary history? Of course not; as Nelson (1978) has shown succinctly, the results of either test merely show one character state, the 'plesiomorphic' one, to be more general than the other (i.e., to occur in more groups)" (1979:543–544).

The problems with the out-group criterion are several. First, determining what the out-group is, a seemingly simple task, can sometimes be virtually impossible (particularly in taxa of which we have no modern phenetic or phyletic understanding). This problem is exacerbated because we never really know at what point the presumptive out-group is attached to the clado-gram of our study group (i.e., at the base or from somewhere within the evolved study group). One can consider different hypotheses of out-group relationships and select states of characters which are apparently primitive in all cases (Coombs, Donoghue, and McGinley 1981), but this can reduce the usable characters to very few, which weakens the basis for the phylo-geny reconstruction. The idea (Watrous and Wheeler 1981) of dividing the study group into evolutionary lines and using Functional In-Groups (FIG) and Functional Out-Groups (FOG) is another helpful approach if no out-group is clearly identifiable. It does allow decisions to be made on polarity of character states. But the polarities determined in this way are based *only* on the few characters and states determined by the broad out-group initially, and such being the case, caution is needed in interpreting the results. Selection of a few other characters and states could change completely the polarities within the study group. The use of FIG and FOG is an in-group concept in which one phyletic line is judged to be the most primitive, giving directionality to all the character states in all of the other phyletic lines.

Second, if an out-group can be identified satisfactorily, than one must be certain of the homologies of the character states, which, because of the abundant parallelisms in angiosperms (see, e.g., in the Compositae, the numerous parallelisms in *Melampodium* [Stuessy 1979] or in *Montanoa* [Funk 1982]), can be very difficult. All one can do is estimate homology based on structural or ontogenetic similarity. The further one goes in search of an out-group, however, as by using higher taxa as out-groups, the greater becomes the probability of having parallelisms in character states. This again emphasizes the related (but basic) idea that attempting cladistic analysis in groups not yet organized phenetically (or phyletically) is less satisfactory than it is in groups in which much preliminary work already has been accomplished.

Third, the difficulty of finding uniform character states is the reason for our using an example in which a nonpervasive, but majority, state occurs (fig. 5.2). This is also the reason we prefer our second stated definition. In practice, it is uncommon in angiosperm systematics to find absolute occurrence of one character state throughout a related group. It is worth pointing out that if two states of a character are found in the study group (hence the systematic interest in the character), it is probable that both states also will be found in close relatives (although perhaps in differing proportions), all having evolved in parallel.

Fourth, we do not possess knowledge about parallelisms or reversals, the absence of which can actually make the out-group criterion misleading. Figures 5.10–5.12 illustrate three hypotheses (cladograms) of evolutionary relationships among four taxa, three in the in-group and one in the out-group (monotypic). Figure 5.10 shows the most parsimonious hypothesis, with a single character-state change having occurred. In-group analysis would erroneously suggest that the dark state is primitive, but out-group comparison clearly shows the light state to be primitive. If one reversal took place in the phylogeny, however (fig. 5.11), the out-group analysis would be misleading but the in-group analysis correct, as would also be the case if one

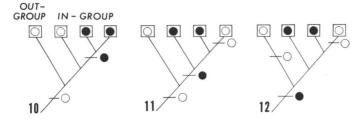

Figures 5.10–5.12. Hypothetical examples of the distribution of a character state, with two states (black and white) in the three extant taxa of the in-group and the one taxon in the out-group (monotypic). 5.10. The most parsimonious cladogram with one derived state change. 5.11. A similar cladogram but with one additional reversal. 5.12. Similar to 5.10 but with one additional parallelism.

parallelism occurred (fig. 5.12). Extinction also may have occurred in many phylogenies, and the analysis of characters and states will probably not reveal this fact, although knowledge of these data would have a strong effect on the reconstruction of the phylogeny. Because we do not know the true phylogeny, we cannot be certain what actually happened. We can (and should) refer to all the other characters, but whether this will resolve the problem depends on their number and nature. We rely largely on parsimony to make our reconstruction of phylogeny. This is fine, but parsimony is clearly not a law of evolution—it is only a methodological prescription (Crisci 1982). Some workers even believe parsimony to have a stultifying influence on the development of ideas in science in general (Dunbar 1980). Almost any cladogram shows some reversals and parallelisms. Parsimony, therefore, can be used as an approximation if no other data exist to the contrary.

Examples can be envisioned in which out-group analysis will be misleading. Consider the hypothetical example in figure 5.13, in which five taxa of the southwestern United States make up the study group and four of these are chromosomally $n=7$ and one $n=6$. The former are found in more mesic habitats and the latter in a more xeric environment. Based on cytogenetic and paleoecological data in comparison with present distributions, it is likely that the $n=6$ taxon was derived from those with $n=7$. Descending aneuploidy in the arid regions of the southwestern United States is a well-documented mode of speciation (Grant 1981). In the related out-group, however, is only one taxon with $n=6$, also found in xeric conditions. The most parsimonious view of the phylogeny using out-group analysis is that $n=6$ and xeric habitat occurrence are primitive for the entire complex. Because of the *nature* of the characters and states (i.e., their evolutionary or biological content), however, it is more likely that this com-

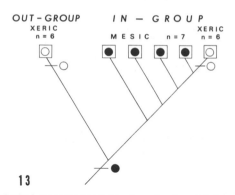

Figure 5.13. Hypothetical example involving two closely related angiosperm taxa of the southwestern United States (in-group with five taxa and out-group with only one), in which habitat occurrence and haploid chromosome numbers are shown. The cladogram illustrated is believed to be most probable (although not most parsimonious), based on the evolutionary (or biological) content of the characters and states.

plex was originally adapted to more mesic conditions and that parallel evolution occurred in both lines as the arid conditions of the Pleistocene developed. This is admittedly a contrived example, but it stresses that caution and the broadest reasoning possible should be entertained before accepting any criterion as providing helpful insights within a particular group. Hypothetical examples involving polyploidy and glaciation could also be given here with equally good effect.

An example in which the out-group criterion is misleading, or at best confusing, comes from *Acmella* L. Rich. (Compositae). This genus contains 30 species and has recently been thoroughly revised (Jansen 1982). The character of interest is the type of flowering head, which exists in radiate or discoid states. Most species (23) in the genus are radiate, fewer (7) being discoid (fig. 5.14). The radiate condition also occurs in all three of the recognized sections (to be validated elsewhere), whereas the discoid condition is confined principally to only two of these three primary evolutionary lines (in two species of section C, occasional discoid populations have been found, but these variants have not been accorded specific status). In-group analysis suggests that the radiate condition is primitive and the discoid condition derived. The closest generic relatives of *Acmella* are *Spilanthes* Jacq. and *Salmea* DC. (Bolick 1981; Jansen 1982). Both of these genera are uniformly discoid. The out-group criterion, therefore, suggests that discoid heads are primitive within *Acmella* and radiate heads derived. This is opposite to the results from in-group analysis.

To resolve the conflicting polarity of head types between in-group and

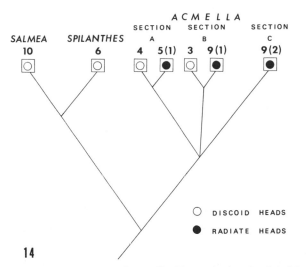

Figure 5.14. Cladogram of sections of *Acmella* (Compositae) and related taxa. Numbers indicate the number of species that have a particular character state; those in parentheses in *Acmella* refer to species that are usually radiate but are known to have occasional discoid populations.

out-group analysis in *Acmella*, other criteria can be examined. If independent information is available that shows evolutionary directionality, then correlation (second-level criterion, fig. 5.8) can be used to give directionality to the different types of heads. Especially meaningful biological information in *Acmella* comes from chromosome numbers, in which diploids, tetraploids, and hexaploids are known (Jansen 1982). From what is understood about cytogenetic mechanisms in angiosperms, the evolutionary origin of hexaploids from diploids and tetraploids is much more probable than the reverse (Stebbins 1980). Of the radiate species in *Acmella*, 13 are diploid, 4 tetraploid, and 2 hexaploid. The discoid species are hexaploid. By correlation, therefore, the radiate head condition in *Acmella* is primitive and the discoid state has evolved in parallel at least three times.

Additional data on breeding systems provide another correlation strengthening the conclusion that radiate heads are primitive in *Acmella*. In the Compositae, five-merous disc florets are the rule and four-merous the exception (Gardner 1977), which strongly suggests that the five-merous condition is primitive. Such a change from five- to four-merous floral structure often is associated with a change in breeding systems from allogamy to autogamy (Ornduff 1969), the latter being regarded as a derived reproductive system. Few data are available for *Acmella* on this point directly, but one discoid species (formerly *Spilanthes ocymifolia* (Lam.) A. H. Moore; the new combination has not yet been made formally) in greenhouse studies is known to be autogamous (Jansen 1982). All the species that have four-merous florets, and are presumptively autogamous, are discoid except for one (formerly *Spilanthes uliginosa* Sw.), which has very small ray florets. These data suggest further that the discoid heads in *Acmella* are derived and the radiate heads primitive.

One might argue that the selection of *Salmea* and *Spilanthes* as the outgroup is inappropriate and that some other taxon with radiate heads should be considered. This could, of course, be true; it only serves to reemphasize the difficulties of selecting an out-group in families that are known at only a rudimentary level (as is the case for most of the angiosperms!). For discussion, if an outgroup to *Acmella* is selected that is taxonomically more inclusive, such as the entire subtribe Verbesininae (sensu Stuessy 1977), both discoid and radiate heads occur, with a preponderance of the latter. However, if we reach even more broadly and consider the entire family as the out-group, then about equal numbers of both types of heads occur, which could make the character weak or inappropriate for polarity determination. The out-group selected, therefore, has a great effect on polarity. The problem in head types discussed here, which is a fundamental characteristic of the reproductive system of the family (radiate to discoid heads represents a change in basic gender strategy from gynomonoecy to hermaphroditism), is that numerous parallelisms have occurred. This is not the exception in the Compositae—it is the rule. All or nearly all characters in the family have undergone extensive parallelisms.

Conclusion

Our main thesis is that caution must be exercised in all aspects of polarity determination. From a general perspective, no aspect of cladistic analysis should be free from careful and thoughtful interpretation, and there are no easy answers or simple solutions. We have no true phylogenies, nor will we ever have them. These considerable difficulties caused many of the pheneticists some two decades ago to reject attempts at phylogeny reconstruction (e.g., Sokal and Sneath 1963; Colless 1967). We believe such efforts are worthwhile, but we also believe that there is no single approach to seeking the true phylogeny. The process of evolution has not been the same in every group of organisms, and therefore the insights to each organismic system will come from different modes of analysis. Criteria of evolutionary directionality are no exception; different criteria are useful depending on the particular group under study. We believe it essential to continue to broaden, rather than narrow, our ideas on assessing polarity of character states. We especially regard the concepts relating to adaptation and function of characters and states as having considerable potential. We also stress the importance of determining the actual modes of speciation by whatever means (especially through experimental crossing studies), to be able to reconstruct phylogeny from the broadest data base possible.

Acknowledgments

Appreciation is expressed to the National Science Foundation for support of the Cladistics Workshop, which enabled T. F. S. to attend the Berkeley meeting; to Daniel J. Crawford and Robert K. Jansen for critically reading the manuscript and making many helpful suggestions; and to the many participants of the Berkeley Workshop who helped shape the ideas presented in this paper.

Literature Cited

Arnold, E. N. 1981. Estimating phylogenies at low taxonomic levels. *Z. zool. Syst. Evolut.-forsch.* 19:1–35.

Bolick, M. R. 1981. A cladistic analysis of *Salmea* DC. (Compositae—Heliantheae). In V. A. Funk and D. R. Brooks; eds., *Advances in Cladistics: Proceedings of the First Meeting of the Willi Hennig Society*, pp. 115–125. New York: New York Botanical Garden.

Camin, J. H. and R. R. Sokal. 1965. A method for deducing branching sequences in phylogeny. *Evolution* 19:311–326.

Cartmill, M. 1981. Hypothesis testing and phylogenetic reconstruction. *Z. zool. Syst. Evolut.-forsch.* 19:73–96.

Colless, D. H. 1967. The phylogenetic fallacy. *Syst. Zool.* 16:289–295.

Coombs, E. A. K., M. J. Donoghue, and R. J. McGinley. 1981. Characters, computers, and cladograms: A review of the Berkeley Cladistics Workshop. *Syst. Bot.* 6:359–372.

Crisci, J. V. 1982. Parsimony in evolutionary theory: Law or methodological prescription? *J. Theor. Biol.* 97:35–41.

Crisci, J. V. and T. F. Stuessy. 1980. Determining primitive character states for phylogenetic reconstruction. *Syst. Bot.* 5:112–135.

Cronquist, A. 1968. *The Evolution and Classification of Flowering Plants.* Boston: Houghton-Mifflin.

——. 1981. *An Integrated System of Classification of Flowering Plants.* New York: Columbia University Press.

Danser, B. H. 1950. A theory of systematics. *Bibl. Biotheor.* 4:115–180.

Darwin, C. 1859. *The Origin of Species.* London: John Murray.

De Jong, R. 1980. Some tools for evolutionary and phylogenetic studies. *Z. zool. Syst. Evolut.-forsch.* 18:1–23.

Delevoryas, T. 1979. Polyploidy in gymnosperms. In W. H. Lewis, ed., *Polyploidy: Biological Relevance,* pp. 215–218. New York: Plenum Press.

Doyle, J. A. 1978. Origin of angiosperms. *Ann. Rev. Ecol. Syst.* 9:365–392.

Dunbar, M. J. 1980. The blunting of Occam's razor, or to hell with parsimony. *Canad. J. Zool.* 58:123–128.

Estabrook, G. F. 1977. Does common equal primitive? *Syst. Bot.* 2:36–42.

Frost, F. H. 1930. Specialization in secondary xylem of dicotyledons. I. Origin of vessels. *Bot. Gaz.* (Crawfordsville) 89:67–94.

Funk, V. A. 1982. The systematics of *Montanoa* (Asteraceae, Heliantheae). *Mem. N.Y. Bot. Gard.* 36:1–133.

Gardner, R. C. 1977. Observations on tetramerous disc florets in the Compositae. *Rhodora* 79:139–146.

Grant, V. 1981. *Plant Speciation.* 2d ed. New York: Columbia University Press.

Hennig, W. 1966. *Phylogenetic Systematics.* D. D. Davis and R. Zangerl, trans. Rpt. 1979. Urbana: University of Illinois Press.

Jansen, R. K. 1982. Systematics of *Acmella* and *Spilanthes* (Compositae, Heliantheae). Ph.D. dissertation, The Ohio State University, Columbus.

Kruskal, J. B. 1956. On the shortest spanning subtree of a graph and the traveling salesman problem. *Proc. Amer. Math. Soc.* 7:48–50.

Mayr, E. 1969. *Principles of Systematic Zoology.* New York: McGraw-Hill.

Nelson, C. H. and G. S. van Horn. 1975. A new simplified method for constructing Wagner networks and the cladistics of *Pentachaeta* (Compositae, Astereae). *Brittonia* 27:362–372.

Ornduff, R. 1969. Reproductive biology in relation to systematics. *Taxon* 18:121–133.

Patterson, C. 1981. Significance of fossils in determining evolutionary relationships. *Ann. Rev. Ecol. Syst.* 12:195–223.

Platnick, N. I. 1979. Philosophy and the transformation of cladistics. *Syst. Zool.* 28:537–546.

Prim, R. C. 1957. Shortest connection networks and some generalizations. *Bell Syst. Tech. J.* 36:1389–1401.

Rieppel, O. 1979. Ontogeny and the recognition of primitive character states. *Z. zool. Syst. Evolut.-forsch.* 17:57–61.

Scadding, S. R. 1981. Do "vestigial organs" provide evidence for evolution? *Evol. Theory* 5:173–176.

Sokal, R. R. and P. H. A. Sneath. 1963. *Numerical Taxonomy.* San Francisco: Freeman.

Stebbins, G. L. 1980. Polyploidy in plants: Unsolved problems and prospects. In W. H. Lewis, ed., *Polyploidy: Biological Relevance,* pp. 495–520. New York: Plenum Press.

Stevens, P. F. 1980. Evolutionary polarity of character states. *Ann. Rev. Ecol. Syst.* 11:333–358.

——. 1981. On ends and means, or how polarity criteria can be assessed. *Syst. Bot.* 6:186–188.

Stuessy, T. F. 1972. Revision of the genus *Melampodium* (Compositae: Heliantheae). *Rhodora* 74:1–70, 161–219.

——. 1977. Heliantheae—systematic review. In V. H. Heywood, J. B. Harborne, and B. L. Turner, eds., *The Biology and Chemistry of the Compositae,* pp. 621–671. London: Academic Press.

——. 1979. Cladistics of *Melampodium* (Compositae). *Taxon* 28:179–195.

Van Valen, L. 1978. A price for progress in paleobiology. *Paleobiol.* 4:210–217.

Wagner, W. H., Jr. 1961. Problems in the classification of ferns. *Recent Advances in Botany* 1:841–844.

Watrous, L. E. and Q. D. Wheeler. 1981. The out-group comparison method of character analysis. *Syst. Zool.* 30:1–11.

Wheeler, Q. D. 1981. The ins and outs of character analysis: A response to Crisci and Stuessy. *Syst. Bot.* 6:297–306.

PART III

Cladogram Construction

INTRODUCTION

With the introduction of the argumentation method by Hennig (1950, 1966) and the Groundplan-Divergence method by Wagner (1961), a new direction for phylogenetic studies emerged. These workers offered explicit methods for the construction of branching patterns of evolution (clado-grams). In addition, Hennig emphasized that the basis for phylogenetic grouping is joint possession of evolutionary novelties.

These methods were expanded and elaborated by various workers, no-tably Edwards and Cavalli-Sforza (1964), who introduced the principle of parsimony as a methodological tool. Camin and Sokal (1965), Kluge and Farris (1969), and Farris (1970) expanded on Wagner's and Hennig's meth-ods and offered numerical implementations utilizing parsimony.

An alternative approach to cladogram construction, compatibility anal-ysis, developed from the ideas of Wilson (1965), Hennig (1950, 1966) and LeQuesne (1969). This method has been refined and discussed in detail by Estabrook and coworkers (e.g., Estabrook 1972, 1978; Estabrook, Johnson, and McMorris 1975, 1976). Compatibility analysis examines the agreement among the estimates of evolutionary relationship shown by each character.

In recent years, Felsenstein (1979) has approached the problem of cladogram construction from the perspective of statistical inference using maximum likelihood models. This approach has demonstrated interrela-tionships among existing methods and emphasized the importance of un-derstanding assumptions about evolutionary rates and the nature of char-acter change to facilitate choosing the appropriate method for a particular group.

The authors in this section discuss the methodologies of parsimony and compatibility, their interrelationships, and their applications. Wagner dis-cusses the applications of his method during the last 25 years. Brooks out-lines the numerical approach to parsimony with a step-by-step guide to the calculation of cladograms using the method of Farris (1970). Estabrook out-lines the relationship between phylogenetic trees and character-state trees as a basis for the compatibility approach. Meacham further pursues com-patibility by examining measures of the probability that sets of compatible characters could occur at random. Felsenstein outlines his maximum like-lihood model of cladistic methods and shows how parsimony and compatibility are related, what assumptions are associated with each, and under what conditions each method would be expected to fail. Baum details applications of both methods to the problem of phylogenetic relationship at different levels in the taxonomic hierarchy, with examples from the grass family (Poaceae). Finally, Fitch stresses a cautious, eclectic approach to determining evolutionary relationships, because of problems associated with all the available methods.

Literature Cited

Camin, J. H. and R. R. Sokal. 1965. A method for deducing branching sequences in phylogeny. *Evolution* 19:311–326.

Edwards, A. W. F. and L. L. Cavalli-Sforza. 1964. Reconstruction of evolutionary trees. In V. H. Heywood and J. McNeill, eds., *Phenetic and Phylogenetic Classification*, pp. 67–76. Systematics Association Publication No. 6. London.

Estabrook, G. F. 1972. Cladistic methodology: A discussion of the theoretical basis for the induction of evolutionary history. *Ann. Rev. Ecol. Syst.* 3:427–456.

———. 1978. Some concepts for the estimation of evolutionary relationships in systematic botany. *Syst. Bot.* 3:146–158.

Estabrook, G. F., C. S. Johnson, and F. R. McMorris. 1975. An idealized concept of the true cladistic character. *Math. Biosci.* 23:263–272.

———. 1976. A mathematical foundation for the analysis of cladistic character compatibility. *Math. Biosci.* 29:181–187.

Farris, J. S. 1970. Methods for computing Wagner Trees. *Syst. Zool.* 19:83–92.

Felsenstein, J. 1979. Alternative methods of phylogenetic inference and their interrelationship. *Syst. Zool.* 28:49–62.

Hennig, W. 1950. *Grundzüge einer Theorie der phylogenetischen Systematik.* Berlin: Deutscher Zentralverlag.

———. 1966. *Phylogenetic Systematics.* D. D. Davis and R. Zangerl, trans. Rpt. 1979. Urbana: University of Illinois Press.

Kluge, A. G. and J. S. Farris. 1969. Quantitative phyletics and the evolution of anurans. *Syst. Zool.* 18:1–32.

Le Quesne, W. J. 1969. A method of selection of characters in numerical taxonomy. *Syst. Zool.* 18:201–205.

Wagner, W. H., Jr. 1961. Problems in the classification of ferns. In D. H. Bailey, Chairman, *Recent Advances in Botany,* 1:841–844. Toronto: University of Toronto Press.

Wilson, E. O. 1965. A consistency test for phylogenies based on contemporaneous species. *Syst. Zool.* 14:214–220.

Section A

Parsimony Methods

6

Applications of the Concepts of Groundplan-Divergence

WARREN H. WAGNER, JR.

The origin and philosophy of the groundplan-divergence method have been outlined recently (Wagner 1980). Its goals are no different from those of phylogenetics in general, namely, the estimation of amounts (grades), directions (clades), and sequences (steps) in evolutionary divergence, using classical ideas of generalized or groundplan character states for judging primitiveness. The steps for doing this, in order, are phenetic classification, detection of hybrids, analysis of character trends, synthesis of generalized character states, estimation of divergence levels, grouping by shared divergence formulas, and connection of clades by hybrid reticulations. The groundplan-divergence method will be referred to in this paper as GPD. Its main objective is to find the simplest phylogenetic explanation of the comparative data.

The GPD method is not concerned with either when or where the evolutionary events took place. It deals only with the amounts, directions, and sequences of the radiating patterns of genetic divergence. These are estimated on the basis of correlations of character states. The same data used in determining branching patterns are used for determining patristic distances. Instead of the actual age of an organism in the fossil record or its geographical distribution, genetic data from the organisms themselves, as evidenced by their phenetic attributes, are used to work out genetic relationships. GPD is also not concerned with formulating taxonomic classifications. Classifications may be a by-product of GPD, but they are not a primary goal. Although chronology, geography, and classification all involve questions that are important in the study of evolutionary relationships, they are nevertheless secondary to the establishment of branching sequences and ancestor-descendant relationships. Our goal is to establish genetic relationships with as high a probability as can be attained. The more parsimonious and compatible the character correlations are, the more probable the resulting phylogenetic trees. If additional knowledge appears that disagrees with the correlations and shows them to be erroneous, the conclusions of GPD are falsified, and the clades need to be revised. The graphical techniques used in GPD combine the cladogram type of presentation used in the argumentation scheme method of Hennig (1966) and the gradogram type of presentation used in the advancement index method of Sporne (1956).

The concepts of GPD are codified in such a way that they can be easily understood and other workers can evaluate the results. It can be used to instruct classroom students in the basic principles underlying systematic biology. As with any good phylogenetic method, the advantages of this one are that it forces the investigator to examine all of the available data and to seek new data as necessary; it rejects questionable character trends that are obscured by confusing multiple parallelisms or by promiscuous and excessive hybridization; it shows what, at the time, is the most probable pattern of phylogenetic relationships; and it "puts the cards on the table," presenting the data and interpretations in such a way that we are enabled to detect defects, accommodate new data, and falsify the conclusions, at least in part.

Today we are faced with many approaches to cladistic and phylogenetic analysis. Many of these are interrelated; the full assessment of their pros and cons has not been achieved. Authors are widely divided as to their opinions on which methods are best. The major disagreements involve how character trends should be interpreted and how cladograms should be used in classification. The underlying theories for determining character trends will be subject to intensive analysis in the near future, so I shall deal with them here only as they have been actually applied to GPD. No doubt, coming years will see greater harmony as a result of better understanding of the logical interconnections of the several procedures currently available. Probably no single method "is always 'best' or gives the 'right' answer" (Duncan, Phillips, and Wagner 1980:291). It is hoped that GPD may be improved in the future by assimilating sound new concepts and more rigorous procedures, whether these are derived from other methods currently available or from new ones.

Applications of GPD over the past 25 years have been diverse (see examples cited in Wagner 1980), some of them quite different from the classroom exercises devised to teach principles of systematics. The method was first computerized by Lellinger (1965). A number of principles in GPD went into the creation of the Wagner Trees by Kluge and Farris (1969), and subsequent modifications and improvements by Farris have produced widely accepted computer programs now available at most large universities. The greatest variation in applications of GPD by different authors has involved approaches to determining trends in character states, as was pointed out in a study by Churchill and Wiley (1980).

The integral role of what I call "groundplan"[1] in GPD was first emphasized in a study of fern classification (Wagner 1961). The principle is followed that character states "found in all or most of a number of related taxa are inherited essentially unchanged from the common ancestor" (p. 843). This principle is based on three assumptions: "*common ancestry*—plants which have in common a majority of similar characteristics have the same common ancestry; *evolutionary divergence*—evolution proceeds normally in various directions, and different lines therefore change in different characters and different character-complexes; and *inequality of evolutionary rates*—evolution occurs at different rates at various times and in different lines. Some forms remain stereotyped and resemble the common ancestor, while others may change radically during the same time period" (p. 842).

There are several pitfalls that must be avoided. The ever-present dangers of misinterpreting reticulate and parallel evolution may be revealed only by general correlations of many characters. The larger the number of characters that are used, therefore, the more accurate are the conclusions (Wagner 1961). Another pitfall is mistaking the most numerous character state as the most primitive. This is caused by the situation in which a divergence may occur in the phylogeny of many clades or on one line that undergoes ex-

---

1. The term "groundplan" is used in different senses, and is being reviewed currently (Wagner, unpublished). Danser defined it as follows: "The systematic groundplan of a natural group is an imaginary living being in which are combined the following qualities: firstly, everything that the groundplans of the component groups of the next lower group have in common; secondly, where these diverge, all the most primitive conditions occurring amongst them" (1950:125). Hennig used the word in a similar sense, writing that "every group formation, irrespective of the rank to which it belongs, must be established by demonstration of derivative ("apomorph") characters in its ground plan" (1965:108). However, the term can be used without phylogenetic implication. For example, in butterfly wing color patterns groundplans (or "archetypes") have been worked out without reference to phylogeny (Nijhout 1981). I define groundplan as used in GPD simply as all of the non-divergent states, taken together, of the common ancestor, hypothetical or real, of a given taxonomic in group. A groundplan in this sense can be determined only by assembling all of the character states that are uniform throughout the in-group and all of the primitive states for characters that have two or more states in the in-group. To determine the latter it is necessary to use comparisons with out-groups (Wagner 1969:81–82, fig. 3). Each derivative subgroup has for its groundplan character states those that are primitive plus the divergent ones that are diagnostic for the subgroup. Only the ancestor for the entire study group including all of its subgroups possesses all of the groundplan character states. The groundplan is not a method but a phylogenetic conclusion based on in-group and out-group character analyses.

cessive speciation, thus making the numerous character state appear to be primitive when it is actually specialized. In general, to counteract this pitfall, data must be obtained from related taxonomic groups at the same level (i.e., through outside comparison). In regard to this, it is important to distinguish between different meanings of the word common, namely, common in the sense of mutual and common in the sense of numerous.

In the application of GPD, often one of the most critical problems involves the establishment of an out-group for comparison. In my research on the fern genus *Diellia* (Wagner 1952), this was the most difficult task, for the genus had been associated taxonomically with several different *families* of ferns (viz., Davalliaceae, Lindsaeaceae, and Aspleniaceae). Determining its true relatives turned out to be the bulk of my study. Once the variables within the genus *Diellia* were established by species-with-species comparisons, their polarities became evident after ascertaining the closest taxonomic relatives in the out-group, the Aspleniaceae.

Different methods of determining polarities are discussed by Crisci and Stuessy (1980) and by Stevens (1980). I question the reliability of Hennig's methods of establishing primitive and advanced states. His four criteria are geological precedence, ecological divergence, ontogenetic precedence, and character correlation (Hennig 1966:95–101). These methods are currently under investigation by several workers. The most reliable procedure is to demonstrate primitive mutuality by determining patterns of character states with respect to related taxonomic groups. Whether we call our comparisons "in-group" or "out-group" depends on the taxonomic size and diversity of the study collection. In a large and diverse study collection, polarities can be determined using the natural partitions. I have discussed outside comparison in connection with GPD in some detail and pointed out that mere "commonness" by itself, in the sense of numerousness, is not sufficient to establish primitive conditions (Wagner 1969: 80–82). I must stress this point, as there has been much misunderstanding regarding it: For comparative purposes, taxonomic distance must be taken into consideration. If one section of a genus contains 8 species, another 35 species, and another only 1 species, the section with only 1 species is equivalent to the one with 35 (Wagner 1980:181). Recent reviews of in-group and out-group comparison methods have been given by Stevens (1980) and Watrous and Wheeler (1981).

The step-by-step description of GPD was first called "Visual Groundplan Correlation Method" (Wagner 1961). The first student to use the method in this form was Mickel (1962) in a monograph of the fern genus *Anemia*. The better name "groundplan-divergence method" was later proposed, to recognize that it utilizes both primitive and derived character states: Successive suites of divergences provide groundplans for each of their respective descendants. The interplay between primitive, or groundplan, character states and derived or divergent, character states provides the framework for establishing the most parsimonious phylogenetic trees. The mutual divergent character-state complexes create the organization of the sequences of taxa. In the early days of GPD, the concepts involved were not nearly as widely accepted by systematists in general, especially zoologists, as they are today.

In the following discussion of applications of GPD, I use to a large extent my own work and that of my students, dealing mainly with higher plants, especially ferns. I also devote attention to various procedural problems that are involved in this and related phylogenetic methods.

Character Trends

Phylogenetics should utilize all known trends to achieve its goals. Functional characters are just as important as structural ones. An ancestor for a given group bears the primitive states not only of morphological characters but of chemical, pollinating, ploidal, phenological, and other characters as well.[2] Shared divergences in all characters, not only uniquely derived but homoplasious ones as well, are necessary for establishing successively more advanced clades, and for these we should adopt not only traditional apomorphous states, but apochemical, apogamous, apoploidal, apophenous, and so on. At the alpha level of GPD analysis, just as in alpha taxonomy, it is true that structural characters will prevail in higher animals and plants, but increasingly more experimental methods will lead to the adoption of more and more biochemical and functional characters, especially in microorganisms. The most meaningful synapotypies should embrace our total knowledge of polarities of characters of all kinds.

In classical biology, many alleged character series were accepted because "they made sense." For example, we had such traditional transformations as simple to complex, indefinite to definite, many to few, like parts to unlike, perennial to annual, woody to herbaceous, radial to bilateral, and

2. It is perhaps unfortunate that the word plesio*morphy* was selected to apply to primitive or groundplan characters. It limits the designation to structural characters. The word *plesiomorphos*, as used traditionally in science, refers to what might be called a morphological convergence, the production of like crystal forms by different compounds. I suggest that a more appropriate root is the combining form basi- (= basal or groundplan). If we substitute that for plesio- (= near), it gives terms like basimorphous, basichemical, and basiploidal to refer, respectively, to primitive morphological, chemical, and ploidal conditions. The popular root apo- is satisfactory in its sense of "away from," i.e., separate, divergent.

so on. Some of these became known to botanists as "dicta," authoritative statements or dogmatic "principles" adopted as bases for phylogenetic trees. We now know that any such generalizations may be fraught with exceptions. In fact, some traditional series have, in certain groups, changed partially or wholly in the opposite direction, producing striking reversals. Reduction in complexity is illustrated over and over again in higher plants—for example, the specialized floating bodies of Lemnaceae or the peculiar flowers of many wind-pollinated trees and graminoids. Amplification from few flower parts to many has also evidently occurred, as in the Cactaceae and Aizoaceae.

For this reason, evaluation of polarities of character trends is a valuable application of GPD for the comparative morphologist. Only by correlating the character in question with many other characters can a morphologist reliably detect reversals or, indeed, any directionalities. That it "makes sense" in ferns that the sporangia progress from simple, tiny leptosporangia to complex, massive eusporangia does not necessarily mean that this pathway is phylogenetically true. The leptosporangium, known only in certain ferns, is in reality a structure peculiar in the picture of all higher plants; in other ferns the eusporangium prevails. More important, all other classes of higher plants outside of the ferns, from trimerophytes to magnoliophytes, possess eusporangia. Eusporangia are thus universal with the exception of certain ferns. This emphatically illustrates that character series and associations cannot be worked out in a disembodied manner. Discoveries of trends can be accomplished only in the light of the entire taxonomic data set.

The same pertains to parallelisms. Since the GPD graph is based on multiple correlations involving numerous characters, it can be a basis for judging what trends represent parallelisms and how many times they reappear. Individual character trends in a GPD analysis may vary from perfect compatibility with other character trends to partial or even complete incompatibility. Parallelism is exceedingly common in higher plants, of course, and the ordinal level in flowering plants supplies many examples, such as reiterated appearances of opposite leaves, herbaceousness, corolla fusion, zygomorphy, and the inferior ovary. In ferns, we see repeated derivation of green, thin-walled spores (Wagner 1974, fig. 4) and reticulate venation patterns (Wagner 1979, fig. 2); thus, as shown in these studies, trilete globose green spores appeared in three widely different groups of ferns and simple-reticulate venation patterns in eleven.

A character analysis that uses phenetic taxonomy to determine trends takes advantage of the maximum information content embodied in the classification and provides a foundation for estimating basi- vs. apocharacters without having to resort to some preconceived idea of a given character series that "makes sense" or depends on a given set of "dicta."

Prediction of Ancestors

Prediction of ancestors has an appeal to scientists in general, because it demonstrates that systematics has a logical structure sufficiently coherent that the researcher can estimate the nature of previously unknown organisms. This happened in the study of the Hawaiian fern genus *Diellia* (Wagner 1952, 1953, 1966). To be able to postulate the course of evolution in a group of organisms on the basis of indirect evidence is one of the goals of phylogenetic research, since in many groups direct evidence is difficult or impossible to obtain. Conclusions are subject to various degrees of confirmation or falsification by the discovery of new evidence.

In general we can only expect to approximate a true ancestor in our predictions on the basis of GPD. The actual discovery of a real ancestor is likely to show that some, but not all, of the character states postulated were correct. Although an ancestor itself may not be discovered, a form divergent from it but closer to it than any other may be found.

For his GPD study of the fern genus *Anemia* subg. *Coptophyllum,* Mickel (1962) investigated the intrageneric relationships of the whole genus *Anemia* and its intergeneric relationships with the sister genus *Mohria*. On the basis of the resulting data he conceptualized a hypothetical ancestor for the whole group that involved 16 groundplan characteristics (Mickel 1962, pl. 10). Subsequently Mickel (1967) found fertile material of a Mexican species previously known only in the sterile state. It confirmed Mickel's postulated course of evolution for the distinctive leaf of *Anemia,* in which fertile basal pinnae are successively more and more exaggerated, erect, and skeletonized. This fern, *Anemia colimensis,* not only shows the most primitive foliar condition known in the genus but shares character states with two of the subgenera, including a number of primitive ones. Although not a perfect ancestor according to the predictions, *A. colimensis* comes closer to it than any other known species and constitutes another illustration of the predictive potentiality of GPD.

In his monograph of Hippocastanaceae, Hardin (1957) did not adopt the GPD method for determining the ancestor but instead elected a taxon, the genus *Billia*, to represent the base. In his concentric graph (Hardin 1957, fig. 15), the primitive type shown at the zero point therefore appears as an ancestor similar to *Billia*. I believe (Wagner 1979), however, that ancestors are best estimated entirely on the basis of the generalized or groundplan character states taken one by one; the ancestor should thus be synthesized.

In studies of certain basidiomycetous fungi, Petersen (1971) also selected existing taxa as ancestors rather than synthesizing them character by character. In estimating the phylogeny of 13 species of *Gomphus*, Petersen chose two different taxonomic groups as possible ancestors, and thus "created" two different groundplans. One of the resultant trees was based on the hypothesis that the ancestor was agaricoid and the other on the hypothesis that the ancestor was like the genus *Clavaridelphus*. If the ancestor were the former, *Gomphus* would be a so-called "degeneration"; if the latter, a "sophistication." Thus, when different roots were chosen, a number of character-state polarities were found to go in opposite directions.

Hauke (1963) in his monograph of *Equisetum* subg. *Hippochaete*, determined the primitive character state on the assumption that "if one [character state] is known to be primitive on other evidence, the others, correlated with it, are also likely to be primitive" (p. 36). This is actually a corollary of the common groundplan principle and is a logical consequence of it. In a given study collection primitive character states tend to coincide more often with other primitive character states than do primitive with divergent states or divergent with other divergent states (unless there is some obligatory functional relationship). This results from the divergence of different clades from the generalized groundplan of their common ancestor and the fact that each clade has its own distinctive advancements that define it. This is a different interpretation of the doctrine of correlation of Sporne (1956; see also below), who proposed that divergent characters would be correlated with divergent characters, not only primitive with primitive.

Plotting the GPD Graph

A suggested nomenclature for GPD graphs is shown in figure 6.1A, using nodes and internodes placed most parsimoniously according to the patterns of common divergent character states and divergence distances. Although the term *straight-line evolution* is sometimes used for ancestor-descendant relationships in a given clade when the ancestor still exists, the internode separating them actually represents a branching. The pathways

Figure 6.1. Methods of constructing GPD and related cladograms. A. Suggested terms for parts of a GPD graph: telome = terminal branch; mesome = any internode between base and telome; basome = unchanged ancestral branch (cf. C_1). B. Different ways of presenting character states: B_1, words for advanced states; B_2, complete divergence formulas, including basicharacters; B_3, only apocharacters; B_4, uniformly aligned character sequence (0 for primitive, 1 for divergent). C. Comparison of different views of GPD: C_1, "from above," showing only estimation of genetic divergence based on shared characters; C_2, "from the side," showing not only divergence but chronology as well (note that the basome itself stops diverging after giving rise to a branch and remains unchanged through time). D. Comparison of a GPD graph for different classifications: D_1, plotted without divergence distance, i.e., all internodes equal; D_2, plotted with DD, two taxonomic units based entirely on monophyly; D_3, plotted

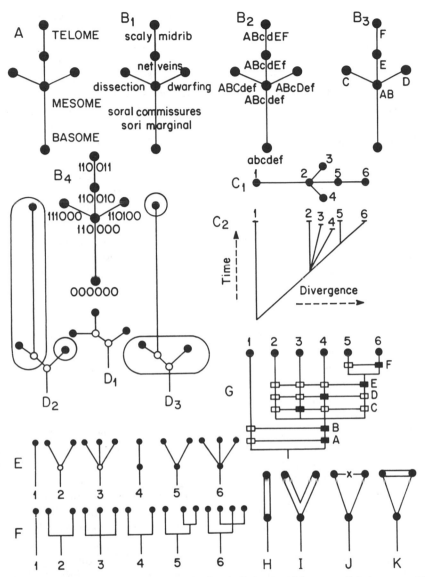

with DD, two taxonomic units based on divergence distance, yielding paraphyletic groups. E, F. Conversion of clade patterns of GPD and HAP, respectively, for various sequences. G. Conversion of GPD graph (as shown, for example, in B_2 and B_3) to a HAP graph (note that the branchings correspond to those in C_2 rather than those in C_1). H. Symbol for a clinal divergence in which there are two extremes, one showing symbasicharacters and one showing synapocharacters, that blend together. I. Symbol like H, but differing in that there are two directions of divergence. J. Symbol for reticulate evolution in which there is a discrete hybrid F_1, sterile, allopolyploid, or apomictic, which does not blend with the parental taxa. K. Symbol for reticulation involving intergradient and blending hybrid swarms.

of genetic change alone are seen when we are "looking down through time" at the tree, but if the tree were seen "from the side" (i.e., if plotted according to a time scale) the persistent ancestor and the descendant would be seen to represent separate branchings, the ancestor retaining its position without further divergence and the descendant undergoing divergent change (fig. $6.1C_1$, C_2). In straight-line evolution there is only a single branching, whereas in dichotomous or trichotomous evolution there are two or three branchings from the same ancestor.

In phylogenetic trees it is conventional to portray all taxa as nodes or points. Often, however, a taxon simply cannot be so tightly defined, because of clinal patterns of variation. If one species has strongly different extremes with a continuum of intermediates between, this may be represented by a "smear" branch, as shown in figure 6.1H, I. Hauke (1963, pl. 11) used the smear symbol to show the clinal ancestor-descendant relationships in one of the subspecies of *Equisetum ramossimum*. Mickel (1962, pl. XI) shows two clinal branches arising within the strongly variable species *Anemia villosa*.

I have found it convenient to use the concentric "bull's-eye" graph, because it presents a readily visualized quantitative estimate of the amount of divergence that has occurred. For comparison of different clades one can read off a scale of divergence values. However, GPD cladograms can also be plotted using internodal length alone (fig. 6.1B).

Divergence Distance (DD) is not, strictly speaking, the same as the Advancement Index (AI) of Sporne (1954), although some authors (e.g., Hardin 1957) equate them. Sporne's AI is based on an interpretation of the doctrine of correlation. It is a percentage derived from a fraction in which the numerator is the number of advanced character states and the denominator is the total number of both advanced and primitive character states. When a taxon possesses both advanced and primitive states of a character, the character is counted as one half. In an example given by Sporne (1954), of a total of 11 characters for Amaranthaceae, 6 are completely advanced and 2 occur in both advanced and primitive states. The resulting fraction, 7/11, gives an AI of 64 percent. Using AIs allows the plotting of gradistic graphs, or gradograms. Groundplans and shared divergent character states for tying together branching sequences are excluded. Clustering is accomplished instead by placing areas representing families on concentric AI circles. Although the families are unconnected to one another, they are grouped radially by ordinal phenetic affinity, as shown in a hypothetical scheme by Sporne (1956, fig. 7). Such gradograms are different from GPD, not only in the way the data are analyzed and the index derived, but in that they es-

chew the use of groundplan altogether. A good recent example of a gradogram is the pattern of angiosperm orders proposed by Stebbins (1974, fig.
11-1).

Divergence Distance (DD) in GPD is simply the total of the character-
state changes as estimated by a groundplan character analysis, giving a value
of 1 for full development of a change or a fraction for partial development.
No effort is made to give greater weight to one character than to another.
The only weighting involves characters for which no trend is evident; these
are eliminated. GPD thus uses the method of weighting by rejection.

Divergence distance is used not only for the construction of trees but also
for estimating their overall length. Also, differences between individual pairs
of taxa can be assessed by totaling their DDs from their common ancestor
if they are on separate branches, or subtracting the smaller divergence from
the larger if they are on the same line.

The hypothetical ancestor has all of the primitive states (basicharacters),
e.g., abcdefghijkl. One clade may have members with suites of derived
characters (apocharacters) ACDGI and ACDGIJ; their DDs are 5 and 6, respectively. Another clade may have members BEF and BEF(L/2); DDs are 3
and 3-1/2. The notation L/2 indicates that a given change is only partially
accomplished, either incompletely developed in all members of the taxon
or present in some members and absent in others. For example, if fully fused
petals represent the completely developed divergence L, then if either of
the aforementioned partial conditions obtains, the intermediate state is given
by L/2. (This should not be confused with hybrid intermediacy, which may
be indicated in the same way but which arises through different processes;
(see Wagner 1980:181, 182, 187; 1983:65–68).

Hall's (1965) Peculiarity Index (PI) is related to both the AI of Sporne and
the DD of Wagner. However, in its derivation it differs from both. In PI the
unusual or peculiar (= divergent) character states for the group are given
values of 1, and the common state 0. The index is derived by taking the
difference between the number of cases of rare states and half the total
number of cases. A state present in only 3 of 40 taxa would be given as
$40/2 - 3 = 17$. Negative weighting for common states is considered super-
fluous. The final PI is the sum of the values for each character. Using 93
characters for 40 taxa of *Eulophia* (Orchidaceae), Hall calculated a range
of PIs from 86 to 261. The relative merits of the three indexes—AI, DD,
and PI—are not known. Each will perhaps have its special advantages for
particular studies involving differing goals. Whatever the case, DD is operationally the simplest of the three and by far the most extensively used
in phylogenetics.

The name "Visual Groundplan Graph," which was originally applied to GPD, suggested only that it illustrates phylogenetic patterns. To be sure, making the pathways of estimated genetic changes readily visible to reader or student is a justifiable goal in any monograph, textbook, or pedagogical exercise. As shown in figure 6.1, the graph may be considered in various ways—"looking down on the tree" or temporally, "from the side," as well as from the tips to the base or the base to the tips. Regarding the last, the graph may be viewed by reading it centrifugally from node to node, so that one encounters more and more complex accumulations of codivergences, or by reading it centripetally toward more and more complete common groundplans, finally arriving at the ancestor. What is primitive or derivative is thus relative to what directionalities are adopted.

In the study of the genus *Diellia* (Wagner 1952, 1953), the ideas of GPD were established in primordial form. A comparison of characters established the affinity of *Diellia* with the genus of the spleenworts, *Asplenium*, and the variables of *Diellia* were then examined for trends, using *Diellia* for in-group comparison and *Asplenium* for out-group comparison. Six trends were found: sori dorsal (a) to marginal (A); sori discrete (b) to commissurate (B); leaflets simple (c) to divided (C); plant size moderate (d) to dwarf (D); distal veins free (e) to netted (E); and rachis glabrous or nearly so (f) to scaly (F). Figure 6.1B shows a parsimonious GPD graph based on these polarities. If the number of variables is few, as in *Diellia,* the divergences can be expressed in words on the cladogram (fig. 6.1B$_1$) or as illustrations of the specialized states (Iltis 1959; Wagner 1966). Both of these methods, especially the latter, are useful for making a visual groundplan tree, but most phylogenetic investigations involve far more than six taxa and six characters. In a roughly "typical" monograph, Evans (1969), working with certain polypodies *(Polypodium),* dealt with 25 species and 23 characters.

Another method of expressing characters is to maintain a uniform sequence (fig. 6.1B$_4$). Thus, on the tree, abcdefghi becomes 000000000. 111000000 gives the divergence formula of ABC, and 110001110 gives the formula ABFGH. This shows the DD readily, but the actual characters are not so easily recognized, especially when the basome is made up of large numbers of characters. It seems to me that the best procedure is to use the letters themselves—small for primitive, capital for divergent (Fig. 6.1B$_2$, B$_3$). It is simple to put down only the new changes, since each node contains the changes of its subjacent mesomes. To obtain the complete divergence formula for a given taxon, one reads from the base to the tip, passing from node to node. The basome itself, of course, calls for no formula, it being understood that all of its characters are at the zero level. One tip, or what

I call a *telome*,[3] *D. mannii*, is expressed only as C, because its immediate ancestor already has AB. Likewise, another telome, *D. falcata*, is characterized by F, since its successive ancestors already have AB and E, in turn.

GPD differs in a number of important respects from Hennig's argumentation plan method (1965, 1966), which I shall here call HAP. The most important involve the manner of expressing ancestor-descendant relationships and the technique of building the cladogram. Regarding the latter, the Hennig method calls for a trellislike structure involving branches, all of which reach the same level; horizontal bars designate characters (fig. 6.1G). One can calculate from HAP the divergence distance of one group from another by determining the total number of either primitive or derivative character states. Even though it seems in HAP that no ancestors are shown, they actually are. HAP calls for sister groups (see below) which tend to change the pattern of branching from that of GPD so that the ancestors in terms of the data are shown as subtending nodes, as shown in figure 6.1G.

GPD and HAP graphs are readily converted one to the other. Figure 6.1E and 6.1F show how the conversions are made. However, GPD is simpler, less "busy." HAP requires more branchings, lines, and symbols to graph the same data (cf. fig. 6.1B$_3$ and 6.1G). I regraphed the GPD of Mickel (1962) using HAP. Mickel used 22 dichotomies to accommodate his estimate of genetic patterns for 41 taxa. Eight of the taxa formed nodes with unidirectional genetic changes, and 3 formed nodes with bidirectional changes. In order to convert the latter to HAP format, I found that 11 additional dichotomies are required, pictorially calling for a 50 percent increase in branchings. Thus, a HAP cladogram may become very complicated indeed.

Hennig's "cladistic principle of dichotomy" requires that every branching create a sister group, no matter what the actual ancestor-descendant relationships are. He believed (Hennig 1966) that every monophyletic group has only one sister group and that, although this is not strictly verifiable empirically, the method has a high heuristic value. The sister-group approach, in his opinion, challenges an investigator to study carefully every case where no dichotomy has yet been demonstrated. In comparison with GPD, use of the sister group does make it somewhat more difficult to visualize ancestor-descendant relationships involved in unidirectional evolution. Even recognizing that the sister-group method is an operational

3. The words *telome* and *mesome* originate from the Telome Theory of Walter Zimmerman of the University of Tübingen, who believed that all branching can ultimately be traced to dichotomy; even apparent "trichotomies" can be resolved into two successive dichotomies. *Telome* is Zimmerman's term for a terminal branch, *mesome* for a nonterminal branch, also called an *internode* (1959:99–101).

convenience for accomplishing certain of Hennig's taxonomic goals, there is the problem, figuratively speaking, that the ancestor becomes a sister taxon of its descendant. The parent and the offspring become sisters; the grandmother becomes a sister of the mother and the mother becomes a sister of the daughter. In this respect, traditional though its graphical approach may be, GPD as viewed "from above" (fig. $6.1C_1$) displays a simpler, less dizzying genealogy.

Some of the criticism of Hennig's sister-group concept is probably semantic and unjustified. The branching pattern is really not different from GPD except that the HAP is graphed with an implied time component (see fig. 6.1G) and all taxa are placed at one time level. In GPD the ancestor or basome represents the conservative ancestral branch which "goes nowhere," i.e., remains nondivergent after dividing into descendant branches. As shown in a side view, with a time component (fig. $6.1C_2$), it stays in the same position as in groundplan-divergence. It may or may not survive until the present; theoretically it or a taxon closely similar to it can exist on the earth at the same time as its descendants. In figure $6.1C_1$, the same basome (taxon 1) appears as seen "from above," i.e., in terms of divergence alone, as a point. The same is true of taxa 2 and 5, which are also ancestors of divergent descendants. Figure $6.1C_2$ shows how the GPD graph in C_1 would look graphed with an implied time component, and comparison of the result with G shows how little the branching patterns differ.

Evans (1969) compared the roles of hypothetical ancestors in different examples of GPD. Pointing out that "where it is necessary to cause a line to fork and there is no extant plant which 'fits' this position, a 'hypothetical ancestor' or 'prototype' to those following it on the lines (indicated by open dots or circles) is inserted" (p. 213), Evans found that one author had to use 19 hypothetical ancestors in evaluating 41 taxa, and another had to use 13 in evaluating only 22 taxa. In his own research, Evans needed only 8 hypothetical ancestors to place 35 taxa. He suggested that a whole genus or subgenus of plants may tend to have a wider spread of character diversity than a single species complex within a genus or subgenus and that the divergence pattern therefore reflects closeness of relationships in such a way that fewer hypothetical ancestors are required. In the future, we may discover that the comparison of the number of hypothetical ancestors with the total number of real taxa in GPD analysis may come to have theoretical value in understanding the nature of the evolutionary processes in different groups. It may be possible to make phylogenetically meaningful comparisons of this phenomenon.

Reticulation

The term *network* or *unrooted tree* has been employed by many authors to refer to cladograms constructed without reference to an ancestor. The classical phenograms of numerical taxonomists may be treated as open, unrooted branch systems, which can be rooted by selecting a taxon, real or hypothetical, and rearranging the branchings so as to arise from that point. In a simple rooted cladogram resulting from an unrooted phenogram, it is not necessary to show branch lengths; all of the branches may have the same length. The mathematical term *network* is now so well established for unrooted trees that it is unlikely to be replaced, whether it is etymologically sound or not. Unfortunately, however, the term in its lay meaning (as well as its traditional use in botany—for example, in venation patterns) refers to a system composed of meshes or areoles, i.e., a reticulum. Unrooted true meshes are well known in a number of plant groups; such networks show origins of taxa by hybridization between normal species (fig. 6.1J, K). They can be constructed in various ways to summarize reticulate phylogeny (e.g., as single triangles, as in *Asplenium* [Wagner 1954, fig. 1]; as several triangles in tandem, as in *Polystichum* [Wagner 1973, fig. 8]; or as successive hybridization loops, as in *Dryopteris* [Wagner 1971, fig. 2]). In order to root such retiform networks it is necessary to work out a GPD graph of the basic divergent components and then make the appropriate hybrid connections. The latter technique was first applied graphically by Mickel (1962, pl. XI), Hauke (1963, pl. 11), and Brown (1964, pl. 1). A simple nonintrogressive hybrid taxon is plotted as showing an intermediate patristic distance between its parents, following the general formula $X = (A + B)/2$. To express the fluctuation of character states in hybrid swarms on a GPD graph, the same use of a "smical" symbol described above in connection with a continuous uni- or bidirectional divergence can be adopted. The two different situations are shown in figures 6.1J and 6.1K, respectively.

Age of Ancestors

Iltis (1959) used a form of GPD in his monograph of *Cleome* (Capparidaceae) and commented, "The new method adds a way of estimating the morphological advancement achieved by the various species in their evolution and thus of placing them in their proper relative positions; this method also groups species according to complexes of shared characteristics. . . . While such diagrams may help us visualize probable phylogenies, . . . morphological specialization is not transferable to mean geological time and/or the age of the species" (p. 131). A GPD graph cannot be used by itself as an estimate of *rates* of evolution, any more than it can state *where* the evolution took place. There is no reason to believe that rates of evolution in one clade are equivalent to rates in another clade. Since GPD estimates only amounts, directions, and sequences of *genetic* changes, the sole way that the speed of change can be determined (in natural populations, as opposed to cultivated plants and domestic animals) is to correlate GPD directly with fossils of known age and sequence. Unfortunately, this is only rarely possible in plant systematics, because of the chance nature of fossil depositions and the rarity of sufficiently complete specimens showing connected parts. A preoccupation with what is called "recency of common ancestry" results, evidently from a belief that the positions of branchings actually do represent the true chronology. All GPD (or any other cladistic technique) can do is estimate the genetic proximities of successive common ancestries. If evolution proceeded slowly, then the prototypic intermediate branchings (mesomes) are ancient; if rapidly, the mesomes are recent. Rates of evolution probably differ not only between clades but on the same clade. Evolution may proceed slowly or stop altogether for long periods, and then suddenly speed up and make many changes in a short spurt. Only fossil evidence can differentiate between antiquity and recency of common ancestry, or between slow and fast rates of evolution.

Places of Origin

Applications of GPD to problems of phytogeography are found in Vitt's 1971 monograph of the moss genus *Orthotrichum* in North America. Using 20 character trends, he worked out a GPD graph of 27 taxa, which he then correlated with their known geographical ranges, and showed that those species with "Arctotertiary" distributions (disjunct in New and Old Worlds) are lower on the clades, on the average, than those that are endemic to North America. The average divergence difference between the two categories of taxa is roughly 20 percent of the maximum divergence found in the genus. For this reason, Vitt concluded that the most specialized forms arose in North America and are more limited in range.

GPD graphs can be plotted directly on maps for an indication of probable migration patterns. However, groups that are very ancient or very stable may perhaps migrate extensively over long periods of time with little change, so that plotting GPD branchings may not be very helpful. Migrations of static, nonevolving forms could occur repeatedly and in various directions in accordance with various climatic changes and geological upheavals. Long-distance dispersal by wind (in plants with minute propagules, such as mosses, ferns, and orchids), by water (in plants with floating propagules), or by birds (in plants with sticky or ingestible fruits and seeds) may also complicate migratory patterns and thus reduce the value of this application of GPD. Vicariance can also obscure dispersal patterns.

Where a group has had a fairly recent origin and active evolutionary change, its migration routes over continents or archipelagos may be estimated using GPD. In his monograph of the New World spurge group, *Euphorbia* sect. *Tithymalopsis,* Huft (1979) was able to plot a topocladogram using GPD, apply it to a map of North America, and thus show areas of probable origins and the directions of migration routes.

Ecological Adaptations

As current biological interest in the significance of ecological adaptation increases, I expect that there will be greater and greater interest in the determination of polarity of character states that are believed to be evolutionary adjustments to different life styles. I do not believe, however, that we should make conclusions of adaptive polarity on the basis of some assumed trend or process based on a dictum or simply on the idea that "it makes sense." Any trend in structure or function that supposedly perfects the ability of an organism to survive in special habitat conditions cannot be accepted until its directionality has been determined first on the basis of GPD. Only after that can we discern the possible adaptive significance, and even at that it is not necessarily true that the trend really does represent an adaptation in the classical sense.

In higher plants many trends have been proposed by botanists, and practically any organ system can be involved in supposed adaptive trends. Among fruit types alone we witness the formation of the samara in connection with wind dispersal, the berry with bird dispersal, nuts with mammal dispersal, and elaiosomes with ant dispersal. In any given group a trend may be reversed, so that grains may become berrylike and berries become nutlike. The parade of diverse desert adaptations includes polarities that are strikingly unlike, representing entirely different modes of adjustment, such as ephemeral life cycles, succulence, and subterranean perennial life cycles.

Closely related taxa having, respectively, the two ends of an evolutionary spectrum of character states can be experimented on to determine whether, in fact, there is any functional support for considering the groundplan character state to have less adaptive advantage than the divergent one. The combination of multiple-character correlation, as in GPD, and experimental confirmation of greater efficiency of the derived state would lend support to a particular hypothesis of adaptation. If the experimental results show the reverse of what would be predicted by GPD, then the hypothesis of adaptive value is falsified, and a new explanation of why the polarity is as it is must be sought.

Especially interesting to both phylogenetics and ecology is the occurrence, so abundantly displayed in higher plants, of parallelism and related phenomena. If the types of correlations made in GPD repeatedly confirm that a given trend undergoes multiple parallelisms, this alone is suggestive of strong adaptive value for the divergent state. An especially disturbing example, however, is the condition of inferior ovary in flowering plants. All of the evidence supports the conclusion that the inferior ovary has arisen dozens of times on many different lines. Yet there is still a real question as to what the true adaptive value or values of the inferior ovary condition may be. We have only speculations such as Grant's (1950) that the inferior ovary condition protects the ovules from birds and beetles. Epigyny still remains an intriguing but poorly understood homoplasy.

Classification

I am not certain to what extent cladistics by itself should modify classification, if at all. Whether monophyly alone should govern classification may be questioned. It is possible, indeed, that most monophyly in the strictest sense can only be guessed at. I believe that phylogenetic trees should be based on classifications, the most logical way of initially ordering the data. Purely phenetic taxonomy does no doubt create paraphyletic groups, and whether divergence distance should enter into classification is unresolved. In this connection, GPD makes it possible to compare classification systems based on branching patterns and sequences of origin by themselves ("pure monophyly") and classifications based on divergence distance (accepting paraphyly, if necessary) as well. An example of this application is shown in figure 6.1D. Mere branching sequences alone will not show this difference clearly (cf. $6.1D_1$ with $6.1D_2$ and $6.1D_3$).

A biological "outsider" today who is not involved in phylogenetic research might gain the impression that systematists, both zoological and botanical, are verging on desperation in their efforts to come up with an acceptable procedure for applying ranks in classification. Hennig's proposal (1966) for solving this problem was that the time of origin of a group should be the basis of ranking higher taxa—the so-called paleontological method. However, even excluding problems of differing personal judgment, the paleontological method of applying categories will probably never work for higher plants or for any group of organisms for which there is only a scanty representation of critical fossils.

Funk and Stuessy (1978) have criticized Hennig's procedure for assigning categories and pointed out the problem caused by differences in rates of evolution in applying it. All taxonomists should, however, state how they arrived at their use of categories, whether by degrees of resemblance between the taxa using strictly phenetic approaches, by cladistic branchings alone, by cladistic branchings plus recency of common ancestor as determined by fossil record, or by some combination of these methods. If some type of character weighting has been adopted, this should also be stated and the criteria for weighting described. Ideally, each monograph should be accompanied by character/taxon divergence tables, discussions for each character used in the study (including the rationale for interpreting its polarity), phenograms, and phylogenetic trees showing sequence of branching as well as patristic distances, i.e., cladistic-gradistic trees, as in GPD. In the future we may give up trying to use taxonomic categories in some consistent way and resort to some other method of classification. It is certainly not clear to me how we can ever achieve stability (Wagner 1969); there is at present too much personal taste and subjective judgment involved, and too little sound theory and objectivity, for making classificatory decisions.

Educational Applications

The groundplan-divergence method was originally designed for use in teaching concepts of systematics. Using it calls attention to the need for critical collections so that broad comparisons can be made using many characters. Actual specimens can be used in the laboratory-classroom, and Cox and Miller (1972) have successfully utilized GPD exercises on flower structure alone, which has not only the advantage of teaching floral organs but that of enabling the student to correlate differences between flower types.

In addition, GPD is of value in showing students that systematics is not concerned solely with the act of identification and naming but that indeed it is a synthetic science that focuses on the *understanding* of diversity. In analyzing character states, students are exposed to the peculiar nature of interspecific and intergeneric hybrids, and they are able to contrast the characteristics of these with normal divergent species. Ideas about the need for a broad array of data and the phenomenon of homology can be developed, as well as such stocks-in-trade of the phylogenist as character analysis, groundplan, divergence, parallelism, convergence, reversal, reduction, amplification, and loss. Various doctrines for estimating ancestors are communicated, especially if the teacher uses actual laboratory exercises and makes it necessary for students to think out the problem themselves. The pleasure of delineating a carefully correlated, nicely parsimonious GPD graph is an intellectually satisfying experience for students and tends to increase their respect for, and appreciation of, systematic science. At the same time, a GPD exercise will tend to emphasize the tentative nature of systematic research in general—what Lincoln Constance (1964) has called the "unending synthesis." GPD gives students the opportunity to evaluate and question the procedures and to decide what evidence, if any, can falsify the conclusions. A student who is primarily a paleontologist can perceive how fossils fit into the picture; a geographer, how places of origin, paths of migration, and processes of dispersal can be hypothesized; a morphologist or physiologist, how polarities can be determined in structure and function; and an ecologist, how the determination of directionalities in character states bears on the evaluation of adaptive significance.

Falsifiable Hypotheses

With GPD systematists try to place the data and correlations in an organized manner enabling other workers to evaluate, modify, and/or reject the conclusions. If our procedures have been correct, other workers can duplicate our results with the same information. This method actually helps to find correlations that otherwise might have been overlooked. The idea that cladograms can be "falsified" is probably, in most cases, too facile, because all degrees of falsification are possible. Character states that show parallelism on different clades do of course "conflict" with "good" character states, those that together all support the same cladogram, but that does not mean that parallelisms should be rejected out of hand or be flatly considered to "falsify" that cladogram. Perhaps they cannot be used to establish its primary framework, but clear-cut parallelisms do represent a substantial portion of the data in phylogenetics and should be added in after the skeleton derived from the "good" characters alone has been established.

Every additional parallelism that accrues in the data set, especially where there are but few exclusively derived sets of attributes, lessens the probability that a given phylogenetic tree is accurate. It does not, however, negate it totally. A good example of large numbers of parallelisms in a tree is shown by Hardin (1957), using 20 characters. Parallelisms occurred in 60 percent of these. Only 8 out of 20 or 40 percent were "good" characters derived only once. In plant taxonomy such large numbers of parallelisms are not at all unusual. Indeed, I estimate from my experience with botanical monographs that percentages from 50 to 80 percent should be expected. Plants are evidently unusually inclined toward parallelisms.

I believe that we lack a theoretical basis for judging how much parallelism is tolerable in constructing GPD graphs. The fewer parallelisms are present, the more likely is the cladogram. If all characters are "good" ones and fit the cladogram without repetition on separate clades, then GPD represents a strongly probable synthesis. The safest procedure is probably to work first only with character states that arise only once and add later those that arise separately twice or more. (To illustrate, compare two cladograms in Wagner 1980: fig. 2, which lacks parallelisms, and fig. 1, which includes parallelisms.) The basic skeleton constructed on the basis of only character states derived once can then "accommodate" secondary cladistic relationships derived from homoplasies. Often, in fact, what we believe to be parallelisms may actually be convergences. Using GPD analysis can help to point these out for further study.

Summary and Conclusions

Students of the history of systematic biology will recognize that there is nothing wholly new or original in the elements of the GPD method. Most of its components are based on time-honored concepts that are as old or nearly as old as evolutionary systematics itself. However, when the different steps involved are clearly enumerated and codified for various applications, the analysis becomes more subject to scrutiny by others, and the way is opened for improvements.

In this article I have discussed some of the applications of GPD in determining character trends, predicting ancestors, assembling shared divergences, plotting cladograms, and dealing with reticulation, geological age, and geographical distribution, and to ecological adaptation, taxonomic classification, and classroom-laboratory teaching. In this light it is to be hoped that in the current spurt of new and different cladistic approaches, the broadly based and traditional procedures of GPD will not be overlooked or lost. If nothing else, they have considerable value in teaching systematic principles, and they lend themselves conveniently to working out phylogenetic patterns in monographic studies. Even today, GPD continues to have a number of advantages over some of the other cladistic procedures, uppermost among which is its simplicity and easy-to-understand logic. Except for extremely bulky problems, involving very high numbers of taxa and character states, GPD can be carried out manually. The data sets and cladograms of GPD can be easily converted to other cladistic formats whenever desired. The gradistic component that is emphasized in GPD is of value for comparing different clades and applications of taxonomic categories. The graphical symbols are designed to handle such special problems as partial changes and/or introgressants. Reticulation is accepted and embodied. Homoplasious trends are given particular emphasis in GPD philosophy in terms of how they are detected and how they can be used in mapping out phylogenetic trees. GPD is subject to various degrees of falsification by reanalysis of individual character states, by discovery of new characters, and by subjection of its conclusions to quantitative estimates of parsimony and character compatibility. In many ways, GPD is already a highly eclectic procedure. Perhaps in the future a fully coherent set of principles and procedures can be synthesized that embraces the entirety of evolutionary theory and embodies the best results of the progress that has been made during the past 50 years.

Acknowledgments

For aid in preparing the manuscript and for many helpful suggestions I wish to acknowledge the following individuals: J. M. Beitel, David M. Johnson, W. Wayt Thomas, Florence S. Wagner, and Kerry S. Walter.

Literature Cited

Brown, D. F. M. 1964. A monographic study of the fern genus *Woodsia*. *Beih. Nova Hedwigia* 16:1–154.

Churchill, S. P. and E. O. Wiley. 1980. A comparison of Wagner's and Hennig's methods of phylogenetic analysis. *Bot. Soc. Amer. Misc. Ser. Publ.* 158:23.

Constance, L. 1964. Systematic botany—an unending synthesis. *Taxon* 13:257–273.

Cox, J. W. and C. N. Miller. 1972. The evolution of flowers: A laboratory exercise. *Sci. Activities* 7:22–25.

Crisci, J. and T. F. Stuessy. 1980. Determining primitive character states for phylogenetic reconstruction. *Syst. Bot.* 5:112–135.

Danser, B. H. 1950. A theory of systematics. *Bibl. Biotheor.* 4:115–180.

Duncan, T., R. B. Phillips, and W. H. Wagner, Jr. 1980. A comparison of branching diagrams derived by various phenetic and cladistic methods. *Syst. Bot.* 5:264–293.

Evans, A. M. 1969. Interspecific relationships in the *Polypodium pectinatum-plumula* complex. *Ann. Missouri Bot. Gard.* 55:193–293.

Funk, V. A. and T. F. Stuessy. 1978. Cladistics for the practicing plant taxonomist. *Syst. Bot.* 3:159–178.

Grant, V. 1950. The protection of the ovules in flowering plants. *Evolution* 4:179–201.

Hall, A. V. 1965. The peculiarity index, a new function for use in numerical taxonomy. *Nature* 205:952.

Hardin, J. W. 1957. A revision of the American Hippocastanaceae. *Brittonia* 9:145–171.

Hauke, R. L. 1963. A taxonomic monograph of the genus *Equisetum* subgenus *Hippochaete*. *Beih. Nova Hedwigia* 8:1–123.

Hennig, W. 1965. Phylogenetic systematics. *Ann. Rev. Entomol.* 10:97–116.

——. 1966. *Phylogenetic Systematics*. D. D. Davis and R. Zangerl, trans. Rpt. 1979. Urbana: University of Illinois Press.

Huft, M. J. 1979. A monograph of *Euphorbia* sect. *Tithymalopsis*. Ph.D. thesis, University of Michigan, Ann Arbor.

Iltis, H. H. 1959. Studies in the Capparidaceae. VI. *Cleome* sect. *Physostemon*: Taxonomy, geography, and evolution. *Brittonia* 11:123–162.

Kluge, A. G. and J. A. Farris. 1969. Quantitative phyletics and the evolution of anurans. *Syst. Zool.* 18:1–32.

Lellinger, D. B. 1965. A quantitative study of generic delimitation in the adiantoid ferns. Ph.D. thesis, University of Michigan, Ann Arbor.

Mickel, J. T. 1962. A monographic study of the fern genus *Anemia* subg. *Coptophyllum*. *Iowa State Univ. J. Sci.* 36:349–383.

——. 1967. The phylogenetic position of *Anemia colimensis*. *Amer. J. Bot.* 54:432–437.

Nijhout. H. F. 1981. The color patterns of butterflies and moths. *Sci. Amer.* 245:140–151.

Petersen, R. H. 1971. Interfamilial relationships in the clavarioid and cantherelloid fungi. In R. H. Petersen, ed., *Evolution in the Higher Basidiomycetes*, pp. 345–371. Knoxville: University of Tennessee Press.

Sporne, K. R. 1954. Statistics and the evolution of the dicotyledons. *Evolution* 7:55–64.

——. 1956. The phylogenetic classification of the angiosperms. *Biol. Rev.* 31:1–29.

Stebbins, G. L. 1974. *Flowering Plants: Evolution Above the Species Level*. Cambridge: Harvard University Press, Belknap Press.

Stevens, P. F. 1980. Evolutionary polarity of character states. *Ann. Rev. Ecol. Syst.* 11:333–358.

Vitt, D. H. 1971. The infrageneric evolution, phylogeny, and taxonomy of the family Orthotrichaceae (Musci) in North America. *Beih. Nova Hedwigia* 21:683–711.

Wagner, W. H., Jr. 1952. The fern genus *Diellia:* Structure, affinities, and taxonomy. University of California Publications in Botany, no. 26.

——. 1953. An *Asplenium* prototype of the genus *Diellia*. *Bull. Torrey Bot. Club* 80:76–94.

——. 1954. Reticulate evolution in the Appalachian aspleniums. *Evolution* 8:103–118.

——. 1961. Problems in the classification of ferns. *Recent Advances in Botany* 1:841–844.

———. 1966. Modern research on evolution in the ferns. In W. A. Jensen and L. K. Kavaljian, eds., *Plant Biology Today: Advances and Challenges,* pp. 164–184. 2d ed. Belmont, Calif.: Wadsworth.

———. 1969. The construction of a classification. In G. Sibley, Chairman, *Systematic Biology,* pp. 67–90. Natl. Acad. Sci. Publ. 1692.

———. 1971. *Evolution of Dryopteris in relation to the Appalachians.* Virginia Polytechnic Institute and State University Research Division Monograph 2, pp. 147–192.

———. 1973. Reticulation of holly ferns *(Polystichum)* in the western United States and adjacent Canada. *Amer. Fern J.* 63:99–115.

———. 1974. Structure of spores in relation to fern phylogeny. *Ann. Missouri Bot. Gard.* 61:332–353.

———. 1979. Reticulate veins in the systematics of modern ferns. *Taxon* 28:87–95.

———. 1980. Origin and philosophy of the Groundplan-Divergence Method of cladistics. *Syst. Bot.* 5:173–193.

———. 1983. Reticulistics. In N. I. Platnick and V. A. Funk, eds., *Advances in Cladistics, Vol. 2: Proceedings of the Second Meeting of the Willi Hennig Society,* pp. 63–79. New York: Columbia University Press.

Watrous, L. E. and Q. D. Wheeler. 1981. The out-group comparison method of character analysis. *Syst. Zool.* 30:1–11.

Zimmerman, W. 1959. *Die Phylogenie der Pflanzen.* Stuttgart: Gustav Fischer Verlag.

7

Quantitative Parsimony

Daniel R. Brooks

This exposition of quantitative parsimony analysis derives its rationale from Kluge (this volume). I do not intend to review the literature or the controversy regarding quantitative parsimony analysis. Rather, I will present the essentials of the method, followed by a brief discussion of the computer program which implements the algorithm (Farris 1970).

Mechanics of Quantitative Parsimony

Step 1

Construct a data matrix (table 7.1). Identify each taxon and each character. For each expression, or *character state*, of each character assign a numerical notation. Plesiomorphic states, as determined by out-group comparisons, are traditionally labeled 0, but any coding convention is feasible as long as the same character state in each taxon is always assigned the same number.

Step 2

Find a root (the out-group X) and connect to it any two taxa (fig. 7.1). All three taxa are joined together at a single point called a *node*. This *three-taxon statement* is called a *Wagner neighborhood*. For any three taxa, there is only one neighborhood possible. The characteristics of the node are defined as the majority state for each binary character or the median state for multistate characters of all characters of the three taxa connected at the node.

Table 7.1. Hypothetical data matrix for sample problem

Taxon	Characters and States											
	1	2	3	4	5	6	7	8	9	10	11	12
X	0	0	0	0	0	0	0	0	0	0	0	0
1	1	0	1	0	0	0	0	1	0	1	0	1
2	0	0	1	1	0	0	0	3	0	1	0	1
3	0	1	1	1	0	0	0	2	1	1	1	1
4	0	0	1	0	1	0	1	0	0	2	1	0
5	0	0	1	0	1	1	2	0	0	2	0	1
6	0	0	1	0	1	1	3	0	0	2	0	1
7	0	0	1	0	1	1	4	0	-1	2	0	1

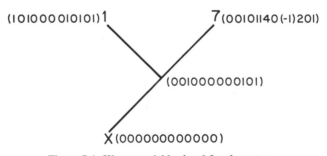

Figure 7.1. Wagner neighborhood for three taxa.

Step 3

Constructing a cladogram using the *Connection Rule* involves searching for the most efficient pattern of shared departures by taxa from common reference points (nodes; see fig. 7.2). This is accomplished by adding taxa one at a time and searching for the most parsimonious connection to the cladogram. This can be done by hand for large numbers of taxa and char-

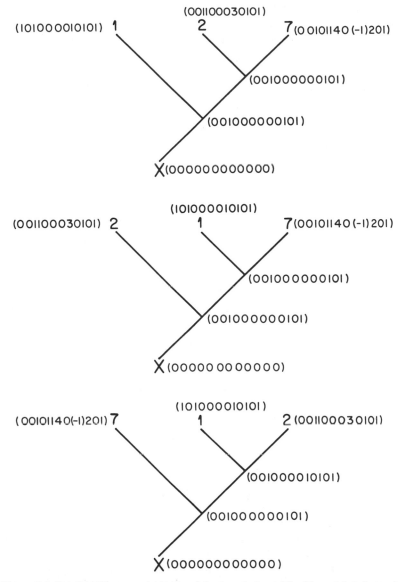

Figure 7.2. Possible Wagner neighborhood for taxa 1, 2, and 7 with postulated ancestor X, using the sample data set in table 7.1.

acters, but becomes laborious if there is much homoplasy (parallelisms and reversals) in the data; in this case, a computer program is needed to perform the calculations. Taxon 2 can be added to the original neighborhood in three possible ways, with the preferred one shown at the bottom (fig. 7.2). One would read the results in the following manner: Taxon 1 and taxon 2 share a unique departure from a reference point which excludes taxon 7. For character 8 the nodal value is the median of 0, 1, and 3, or 1. Taxon 3 can now be added to the cladogram and the preferred node designated (fig. 7.3).

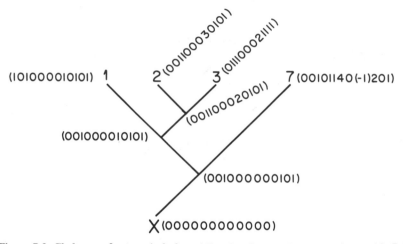

Figure 7.3. Cladogram for taxa 1, 2, 3, and 7, using the sample data set from table 7.1.

Step 4

If taxon 4 is now added to the cladogram, there are seven possible connections, five of which are distinct. All five of the connections and their calculated nodal values are shown in figure 7.4. The parsimony criterion is invoked to choose the preferred cladogram. Notice that only two of the five postulated distinct nodes are actually new; the remaining three have been calculated previously for other taxa. The total number of differences in character states between the node and taxon 4 for cladogram A is eight; for cladogram B, seven; for cladogram C, five; for cladogram D, seven; and for cladogram E, two. Nodal character-state values that either maximize the number of shared departure points or minimize the number of postulated new character-state changes are preferred. Thus, cladogram E, on which a

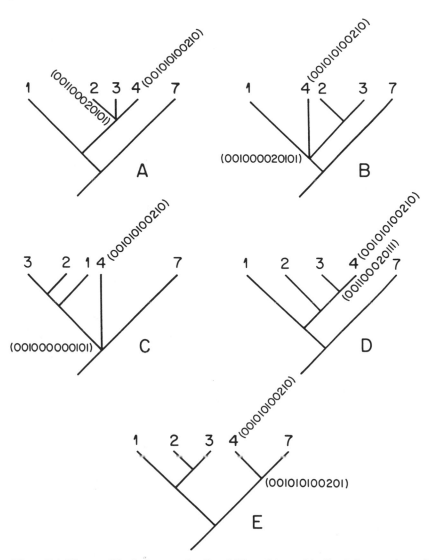

Figure 7.4. Five possible cladograms for the addition of taxon 4 to the cladogram shown in figure 7.3, using the sample data set from table 7.1.

new node is postulated and which requires only two character-state changes, is preferred. With this node requiring only two changes and a previously calculated node requiring only five changes, we postulate that the trait linking taxon 4 and taxon 3 (character state 1 for character 11) is the result of convergence rather than common ancestry.

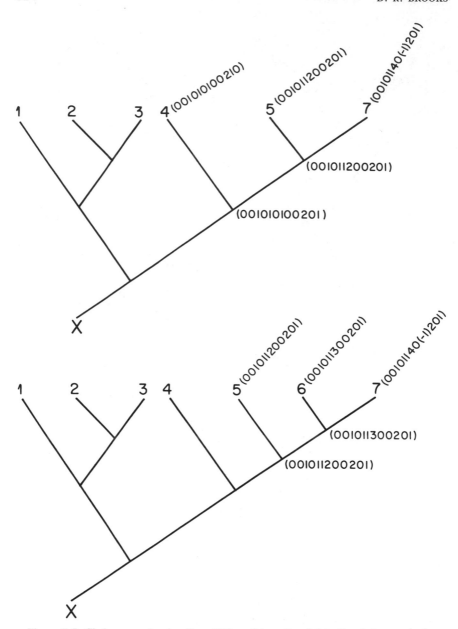

Figure 7.5. Cladograms showing the addition of taxa 5 and 6 to the cladogram in figure 7.4, using the sample data set from table 7.1.

Step 5

Figure 7.5 shows the addition of taxa 5 and 6 to the cladogram and the nodal values calculated for the most parsimonious connections.

Step 6

Remove the root, X, and add numerical shorthand notations indicating shared derived character states (*synapomorphies;* see fig. 7.6). Each slash mark indicates a synapomorphic trait for the character denoted by the accompanying number; a cross indicates an apomorphic trait denoted by a negative sign (e.g., -1 for character 9); and an asterisk indicates a postulated reversal (e.g., 0 for character 12 in taxon 4). Apomorphic traits for multistate characters are determined by summing the slashes for a given character from the bottom of the cladogram. Thus, the "7 slash" on the branch leading to the cluster taxon 6 + taxon 7 represents state 3 for character 7.

Step 7

Multistate characters pose some problems for phylogenetic analysis. First, the greater the number of states, the greater the difficulty in determining the primitive state correctly, at least initially. Second, it has been suggested

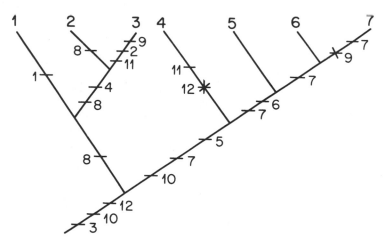

Figure 7.6. Cladogram showing types of character changes leading to the taxa from the sample data set in table 7.1. See text for explanation of notation.

that multistate characters unduly influence the results of a cladistic analysis based on data that are mostly binary. Third, complex multistate characters cannot be coded directly for computer-assisted quantitative parsimony analyses. The first two objections may be overcome by using the "Wagner" algorithm (Farris 1970) rather than a more restrictive technique like the Camin-Sokal method. All three problems may be overcome if all multistate characters are converted into series of binary characters. The technique for such comparisons is called *Additive Binary Coding.*

Consider a complex multistate character, with nine states related to each other, as shown in figure 7.7. To convert this *character-state tree* into a set of binary characters, construct a state-by-state matrix (table 7.2). Then, be-

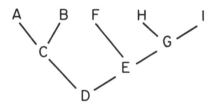

Figure 7.7. Character-state tree for hypothetical character with nine states.

ginning with row A, enter the value 1 in each column representing a state linking A with the basal D, inclusive. For A that would be A, C, and D. For row B, columns B, C, and D would be scored 1. All slots in the matrix not scored with a 1 would be filled with 0. The final matrix comprises a set of columns each representing a character state from the original tree and a set of rows each corresponding to a new binary character. The total data matrix is a binary representation of the entire character-state tree.

This technique adds several dimensions to phylogenetic analysis. First, it allows precise formulation of median values for multistate characters. As an example, consider the Wagner neighborhood for three taxa characterized by states A, E, and H fom the sample binary matrix. What value would be placed at the node for such a neighborhood? The solution to the problem involves designating A, E, and H with their binary coding and finding the median values of the binary characters. The answer is 000110000, or E. This technique also serves a critical function in the analysis of coevolutionary relationships among symbionts (Brooks, 1981).

Step 8

Optimize the tree. This technique provides a way to derive parsimonious inferences about phylogenetic sequences of character-state changes from a

Book by
Wiley on Calculus
good for learning about

Table 7.2. Conversion of data of single character-state tree
(fig. 7.7) into nine binary characters

New Binary Characters	Old Character States								
	A	B	C	D	E	F	G	H	I
A	1	0	1	1	0	0	0	0	0
B	0	1	1	1	0	0	0	0	0
C	0	0	1	1	0	0	0	0	0
D	0	0	0	1	0	0	0	0	0
E	0	0	0	1	1	0	0	0	0
F	0	0	0	1	1	1	0	0	0
G	0	0	0	1	1	0	1	0	0
H	0	0	0	1	1	0	1	1	0
I	0	0	0	1	1	0	1	0	1

tree. Consider the tree in figure 7.8. Notations at the ends of branches refer
to presence (1) or absence of (0) of a state. The question in optimization is
this: What is the most parsimonious interpretation of the evolution of those
character states? The solution involves a two-step process. First, pass from
the top of the tree to the bottom, assigning values to the nodes according
to the following rules: If the two branches running from above down to the
node have the same character state (0 or 1), code the node the same way;
if the character states differ (one 0 and one 1), assign the node a b for both.

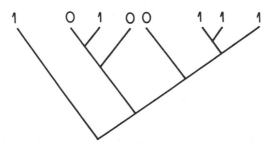

**Figure 7.8. Cladogram labeled for presence or absence of a state of a character in the taxa
that occupy these points for use in tree optimization.**

If a b and a 0 come together, assign the node a 0; if a b and a 1 come together, assign the node a 1; for two b branches, assign the node a b. Figure 7.9 shows the properly coded nodes. The second step in optimization requires designating all b nodes as either 0 or 1. The most parsimonious designation of each b is the same value as that exhibited by the outgroup (the previous nodal value). The solution is shown in figure 7.10. The most parsimonious explanation of the sequence of character-state changes is that state 1 evolved into 0 at x and then reappeared twice, once at y and once at z.

Optimization also provides means for determining which of two or more trees best fits a set of data. Consider an alternative tree (fig. 7.11). Optimizing this tree produces an interpretation that state 1 evolved into state 0 once. The total number of postulated changes for the tree in figure 7.8 is three; for the tree in figure 7.11 it is one. Thus, the second tree represents

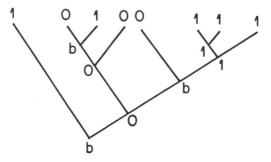

Figure 7.9. Cladogram with interior nodes labeled according to the tree optimization procedure discussed in text.

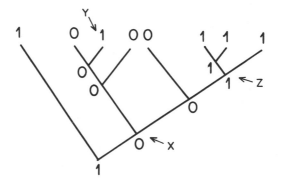

Figure 7.10. Cladogram with interior nodes labeled b in fig. 7.9 relabeled as 0 or 1, according to the tree optimization procedure discussed in text.

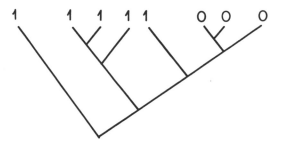

Figure 7.11. Alternative cladogram for comparison of trees, using the tree optimization procedure discussed in text.

a more parsimonious explanation of the character data. Such comparisons can be made for all characters in a data set in terms of any possible alternative tree.

Computer Implementation

The Wagner-78 computer program designed by Dr. James S. Farris to implement research programs based on quantitative parsimony analysis has three primary capabilities. First, it makes it possible to derive cladograms from matrices of distances using the Wagner distance routine (see Farris 1972 for a description of the technique and Farris 1981 for a discussion of the validity of distance measures in phylogenetic analysis). Second, it can generate a goodness-of-fit statistic for any cladogram over any set of data using the Tree routine. This routine allows any tree produced by any means to be compared directly with any other trees so that a preference for one may be expressed based on observation of how well each of them fits the data at hand. Third, the main subroutine calculates a Wagner Tree based on a matrix of characters, either discrete characters or continuous data variables. One may also select a number of possible additional output data, such as apomorphy lists or character consistency ratios, to be printed along with the tree. In order to reduce computer time and expense, Farris' program includes a computational shortcut. The data matrix is converted to a Manhattan Distance Matrix, a Prim Network is constructed from the matrix, and the network is then optimized (see section on optimization) to produce the most parsimonious cladogram.

Manhattan Distance may be most easily understood as the number of character-state differences between any two taxa divided by the total number of characters. Table 7.3 depicts the Manhattan Distance Matrix for the

Table 7.3.   Manhattan Distance Matrix for data in Table 7.1

Taxon	Taxon							
	X	1	2	3	4	5	6	7
X	–	0.42	0.58	0.75	0.50	0.67	0.75	0.92
1	0.42	–	0.33	0.50	0.58	0.58	0.67	0.83
2	0.58	0.33	–	0.33	0.67	0.75	0.83	1.00
3	0.75	0.50	0.33	–	0.75	0.92	1.00	1.17
4	0.50	0.58	0.67	0.75	–	0.33	0.42	0.58
5	0.67	0.58	0.75	0.92	0.33	–	0.08	0.25
6	0.75	0.67	0.83	1.00	0.42	0.08	–	0.17
7	0.92	0.83	1.00	1.17	0.58	0.25	0.17	–

data in table 7.1. The closest neighbor(s) for each taxon based on the Manhattan Distance values (table 7.3) is found as follows: closest to X = taxon 1; to 1 = 2, 3; to 2 = 1, 3; to 3 = 2; to 4 = 5; to 5 = 6; 6 = 5; to 7 = 6. Figure 7.12 depicts the Prim Network for the above. That network is then rooted by the out-group (X, the ancestor or the first line of data read into the program; see fig. 7.13), and then optimized (fig. 7.14) to give the final cladogram.

I believe the major reason such a small number of systematists have investigated the utility of quantitative parsimony techniques is the perception that such techniques require extensive knowledge of complex mathematics. It is true that Farris and others have provided extensive and erudite

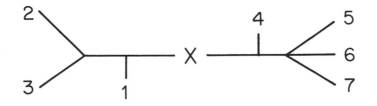

Figure 7.12. Prim Network for the taxa in table 7.1, based on the Manhattan Distance separating them.

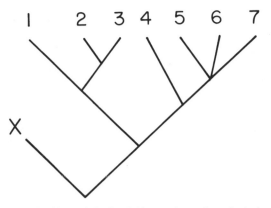

Figure 7.13. Prim Network in fig. 7.12 rooted at a hypothetical ancestor.

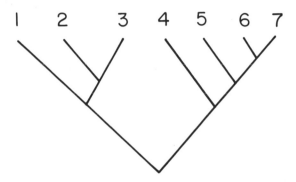

Figure 7.14. Cladogram in fig. 7.13 optimized according to procedure outlined in text.

documentation of the mathematical bases of and justification for quantitative parsimony techniques. But, as I hope I have demonstrated, the techniques represent only a logical extension of the concepts of phylogenetic systematics first elaborated by Hennig (1966) and explicated so well in Kluge's paper in this volume. As such, their mechanics can be applied by people who do not possess special mathematical skills or interests.

There have been almost 25 years of sentiment in systematics that we ought to doubt our sense perceptions and rely on "objective" constructs for analysis. One collects a body of data without any consideration of its qualities, and subjects this to a variety of analyses until one gets the "correct" answer (a more extreme position advocates that one simply publish all results without any scientific judgment, then move on). The problem with that attitude is that we have no way of acquiring knowledge, i.e., gauging the verisimilitude of theories, or of judging sense from nonsense without using our sense perceptions. To that end, phylogenetic systematics is the logical protocol for systematists. One may perform a cladistic analysis on a set of data and produce nonsense. At that point, the onus is not on the technique, which represents an accurate summation of the data set, but on the data themselves and, by extension, on the scientist. Phylogenetics directs systematic research toward more and better character analysis rather than toward more clustering algorithms and jazzier computer programs.

Literature Cited

Brooks, D. R. 1981. Hennig's parasitological method: A proposed solution. *Syst. Zool.* 30:229–249.

Farris, J. S. 1970. Methods for computing Wagner trees. *Syst. Zool.* 19:83–92.

——. 1972. Estimating phylogenetic trees from distance matrices. *Amer. Natur.* 106:645–668.

——. 1981. Distance data in phylogenetic analysis. In V. A. Funk and D. R. Brooks, eds., *Advances in Cladistics: Proceedings of the First Meeting of the Willi Hennig Society*, pp. 3–23. New York: New York Botanical Garden.

Hennig, W. 1966. D. D. Davis and R. Zangerl, trans. Rpt. 1979. *Phylogenetic Systematics*. Urbana: University of Illinois Press.

Section B

Compatibility Methods

8

Phylogenetic Trees and Character-State Trees

GEORGE F. ESTABROOK

Phylogenetic Trees

One common approach to phylogenetic reconstruction is to structure comparative data into several character-state trees, and then somehow to combine these into a phylogenetic reconstruction. Here I discuss the relationship between phylogenetic trees and character-state trees and describe how character-state trees may be construed as hypotheses of phylogenetic relationship.

Fundamental to all that follows are these assumptions: species have somewhat more biological reality than other taxa; each species has evolved from a single immediately ancestral species; and each species differs from this immediate ancestor in some properties that change during this evolution.

A phylogenetic tree for some collection S of species under study is a tree diagram representing the genealogical continuity of species through time. Each branch point in the tree corresponds to a speciation event, and other points may correspond to speciation events as well. The number of lines at any given time is the number of species at that time. No species is represented at any single time by more than one line.

The phylogenetic tree of figure 8.1 represents the phylogenetic lines below the species: a, b, c, d, e, f, and g. These species, together with their ancestors, correspond to various line segments of this phylogenetic tree. If the arrows in figure 8.2 point to all the speciations that occurred, then there would be only these seven species in this phylogenetic tree, and all their ancestors would already be among them. The right side of figure 8.2 shows that the Haase diagram of the ancestor-descendant relationships among these

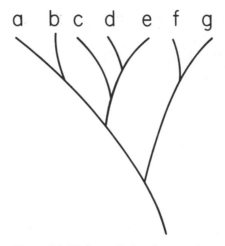

Figure 8.1. Phylogenetic tree for seven taxa.

seven species would be in this case. Such a diagram shows ancestor-descendant relationships by connecting, with lines going toward the top of the diagram, each ancestor to each of its immediate descendants. Here ancestors and descendants are all species, but Haase diagrams may also be drawn in which ancestors or descendants are groups of species.

The arrows in figure 8.3 show another possibility for speciations in the

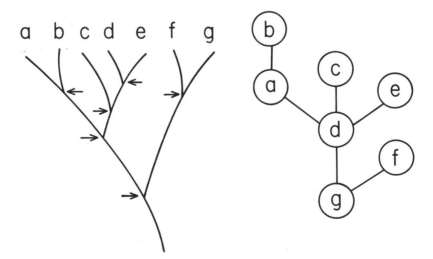

Figure 8.2. Phylogenetic tree from fig. 8.1, with one possible set of speciation events marked with arrows, and the resultant Haase diagram.

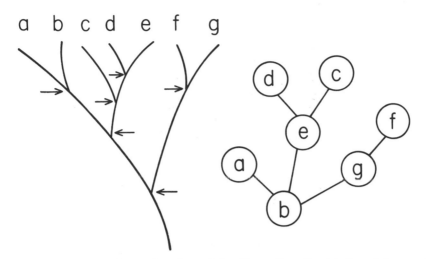

Figure 8.3. Phylogenetic tree from figure 8.1, with an alternative set of speciation events shown with arrows, and the resultant Haase diagram.

phylogenetic tree of figure 8.1. The right side of figure 8.3 is the Haase diagram of ancestor-descendant relationships among the species for this case. Notice how different the ancestor-descendant relationships can be for the same species with the same phylogenetic tree, but with different patterns of changes in their histories.

There may be more than seven species in this phylogenetic tree. The arrows in figure 8.4 represent the same speciation events as shown in figure 8.3, but an additional speciation, shown by the star, has been included. The resulting Haase diagram is shown on the right, where the empty circle indicates an ancestor that is not represented in the study collection. One possible description of the events at "star" is that during the speciation of *b* this ancestor became extinct. But other possibilities exist: the extinction of this ancestor occurred later; or this ancestor is not extinct, just not represented in the study. Figure 8.5 represents with *x* such a species. Because the phylogenetic tree for a study collection S of species contains only the phylogenetic lines under the species in S, many such "branches" to extinct or unobserved species may exist.

To test your understanding, cover the right half of figure 8.6 and try to draw the Haase diagram of the ancestor-descendant relationships implied by the nine numbered arrows indicating speciation events on the phylogenetic tree of figure 8.1. The right half of figure 8.6 is that Haase diagram. Notice that the lines in the Haase diagram correspond to speciation events

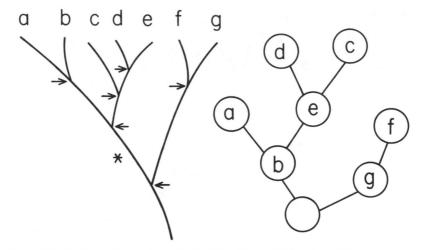

Figure 8.4. Phylogenetic tree shown in fig. 8.3, with an additional speciation event marked with an asterisk, and the resultant Haase diagram.

in the phylogenetic tree, as shown by the numbering. In all the above examples, the phylogenetic trees have been the same, but the speciation events have been different and therefore the ancestor-descendant relationships have also been different. Thus, the same phylogenetic tree can correspond to many different ancestor-descendant relationships for the same group of species,

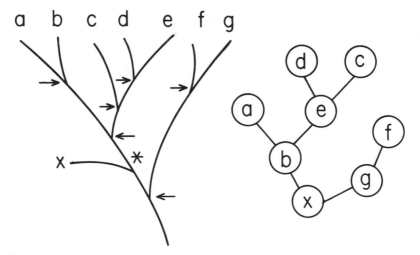

Figure 8.5. Phylogenetic tree shown in fig. 8.4, with ancestral species *x* added to the Haase diagram.

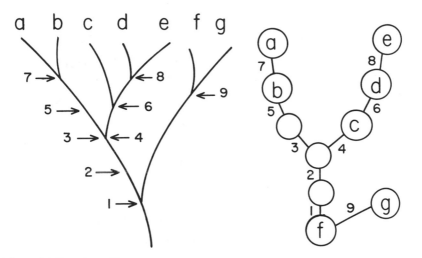

Figure 8.6. Exercise to illustrate the relationship between phylogenetic trees and Haase diagrams.

depending on what speciation events happened in the past. As we shall see, however, some pairs of Haase diagrams of hypothesized ancestor-descendant relationships for the same study collection of species cannot both correspond to the same phylogenetic tree.

Character-State Trees

A *character-state tree* for a study collection S of species is

a) a dividing of S into groups of species, called character states, so that each species belongs to exactly one character state, and

b) a tree Haase diagram including these character states and possibly some additional "empty" states, subject to the restriction that every empty state have at least two immediately descendant states.

This concept is illustrated in figure 8.7. In character-state tree i, S has been divided into three states *(ab, cde,* and *fg),* and there are no empty states in the Haase diagram. In character-state tree ii, an empty state is included with these same states. In the Haase diagram this empty state has two immediate descendants: *ab* and *cde.* In character-state tree iii, three empty states are included, but one has no descendants and one has only one descendant, state *ab.* These two empty states must be eliminated, re-

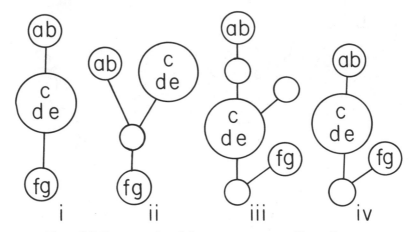

Figure 8.7. Four examples of character-state trees as Haase diagrams.

sulting in character-state tree iv. The reason for the elimination of these empty states is to simplify the concepts to follow and to make possible a very useful correspondence. In addition, you will see that rarely if ever is information available with which to include empty states with fewer than two descendants.

Haase diagrams of ancestor-descendant relationships among species are special kinds of character-state trees in which each species in S is in a state by itself and the empty states correspond to unobserved ancestors. It is entirely possible for some or all of the species in S to be extinct, as long as specimens (fossil or otherwise) are at hand to provide an observational basis for their comparison and description. One useful way to view character-state trees is to consider them to be fully or partially resolved Haase diagrams of ancestor-descendant relationships among species. In the case of partially resolved diagrams, character-state trees fail to distinguish among all the species because they omit some of the speciation events.

Speciations on the phylogenetic tree of figure 8.1 that correspond to the character-state trees of figure 8.7 are shown in figures 8.8 and 8.9. For example, the speciation corresponding to the line below character state *ab* is shown by the upper arrow in figure 8.8. Breaking the phylogenetic tree in this arrow separates species *a* and *b* from the rest. In addition, none of the speciations that distinguish *c, d,* or *e* from each other or from their immediate ancestors is shown except for the speciation that determines their character state in character-state tree i. In figure 8.9, character-state trees ii and iv correspond to the inclusion of one additional speciation event, in a different place for each, to give rise to a different empty state in each case. Note that each empty state has two descendant states.

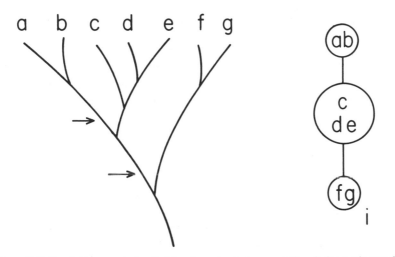

Figure 8.8. Speciation events implied by character i shown on the phylogenetic tree from fig. 8.1.

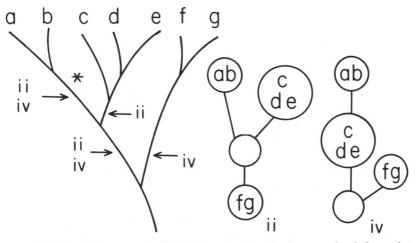

Figure 8.9. Speciation events implied by character ii and iv shown on the phylogenetic tree from fig. 8.1.

Character-State Trees as Hypotheses

For a character-state tree to be a hypothesis, it is sufficient to specify the conditions under which it would be true and the conditions under which it would be false, so that the possibilities are exhausted. According to the fundamental assumptions set forth above, there is one true phylogenetic tree, and if alpha taxonomy has been done correctly so that the species recognized are actually separated by speciation events, then one of the many possible patterns of speciations and distinctions among the ancestors actually occurred. A character-state tree is true if it is the Haase diagram of ancestor-descendant relations derived from the true phylogenetic tree through consideration of some of the speciations that actually occurred.

All the character-state trees that have been shown so far could be true for the phylogenetic tree of figure 8.1, provided the "right" speciations occurred. Figure 8.10 illustrates some additional character-state trees that could be true for this phylogenetic tree. The one-state character is always true for any phylogenetic tree. A character-state tree with only a single species in its only advanced state can always be true for any phylogenetic tree, given the appropriate speciation events, but in fact may not always be true. The same is true for the "fan"-shaped character-state tree in the upper part of figure 8.10. Unless a character-state tree hypothesizes a speciation event common to the ancestry of two or more species in S, it can (given the appropriate speciation events) be true for any phylogenetic tree.

Some character-state trees that cannot be true for the phylogenetic tree of figure 8.1 (no matter what speciation events may have occurred on it) are illustrated in figure 8.11. No possible placement of speciations on the phylogenetic tree can give rise to these character-state trees. If the phylogenetic tree of figure 8.1 is true, then all these character-state trees are false.

There are three basic ways in which character-state trees can be false. The most severe is "states," which occurs when there is no pattern of speciation events that could produce the states of the character. If the states of a character-state tree can be produced by placing speciations on the true phylogenetic tree but the Haase diagram is wrong no matter which such pattern is used, then the problem is "trends" or "proximity." A trends problem can be corrected by a simple redirecting of the edges of the character-state tree (the lines connecting the states). This is illustrated in figure 8.12 where state *fg* is made primitive (instead of state *cde*), thus reversing the evolutionary trend in which state *cde* gives rise to state *fg*. If a character-state tree does not have a "states" problem but it cannot be corrected by reversing some trends, then there is a "proximity" problem; some edges of the character-state tree must be broken and new ones established, as illustrated in figure 8.13.

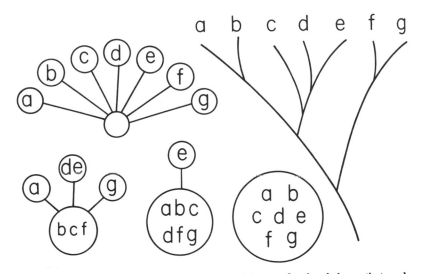

Figure 8.10. Various character-state trees that could be true for the phylogenetic tree shown to the right.

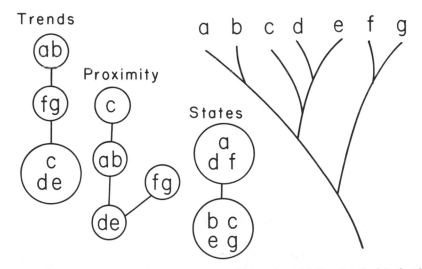

Figure 8.11. Examples of character-state trees and ways in which they may be false for the phylogenetic tree to the right.

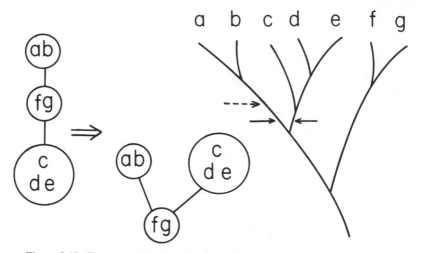

Figure 8.12. Character-state tree showing a "states" problem and its resolution.

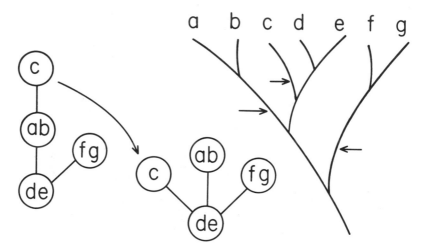

Figure 8.13. Character-state tree showing a "proximity" problem and its resolution.

Addition of True Character-State Trees

A true character-state tree indicates some of the places on the phylogenetic tree where speciations occurred. The addition of two true character-state trees gives a third true character-state tree, which indicates all the speciation places from the two trees being added. Figure 8.14 shows the addition of tree i and tree ii to get tree iii. Arrows labeled i show the speciation places indicated by character-state tree i, and the arrow labeled ii shows the speciation place indicated by character-state tree ii. Character-state tree iii, their sum, contains all three of these speciation places.

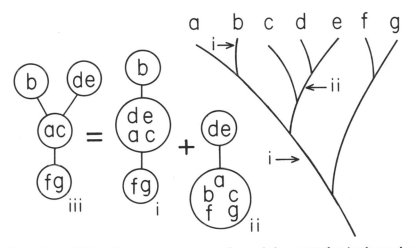

Figure 8.14. Addition of character-state trees; the speciation events they imply are shown on the phylogenetic tree to the right.

Character-state tree iii (the sum of i and ii) can in turn be added to another true character-state tree as shown in figure 8.15. The same labeling convention is used to indicate how character-state tree v is determined. We see that the sum of iii and iv generates an empty state where neither term had one. We also see that the sum includes two speciations on the internode of the phylogenetic tree leading to species c and d. There are several possibilities to account for this: the speciation indicated by iii may be the same as that indicated by iv; iii may have preceded iv; or iv may have preceded iii. In the last two cases an empty state would be generated in the sum, but in the first case no empty state would be generated. Because there is no way to distinguish among these possibilities on the basis of character-state tree information, I choose to simplify matters by excluding any empty state that might otherwise be generated by this situation. Such empty states would have only one immediate descendant; this provides further explanation for constraint b in the definition of character-state trees.

It is important to realize that what can be constructed from the simultaneous consideration of true character-state trees is always another charac-

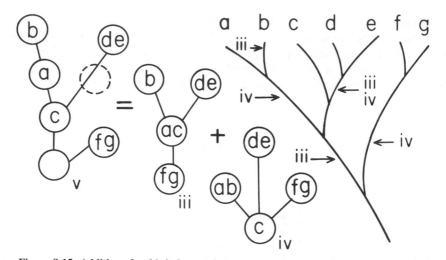

Figure 8.15. Addition of a third character-state tree to the two in fig. 8.14; the speciation events implied are shown on the phylogenetic tree to the right.

ter-state tree. In particular, the phylogenetic tree itself cannot be recovered in every detail from true character-state trees. What can be constructed from true character-state trees must have the form of a character-state tree. The information in the phylogenetic tree that can be recovered with character-state trees is the information given in the very refined character-state tree that is the Haase diagram of ancestor-descendant relationships among all the species, determined by consideration of all the speciation events that actually occurred on the lines of the phylogenetic tree. The most refined character-state tree possible for a phylogenetic tree occurs when there is at least one speciation per internode; but this possibility has not necessarily been realized, and there may have been fewer speciations in the evolutionary history of the group (a situation that would not necessarily invalidate the alpha taxonomy).

An alternative, equivalent representation of the information in this most refined character-state tree (if indeed it even existed as a historical reality) is the list of all subsets of S maximally contained in monophyletic groups (groups of species containing every species that is a descendant of the group's most recent common ancestor). Supposing at least one speciation per internode as before, such a collection of subsets of S is shown in figure 8.16.

Observe that S, the whole collection of species under study, is a subset on this list. In addition, any two subsets on the list are either exclusive or

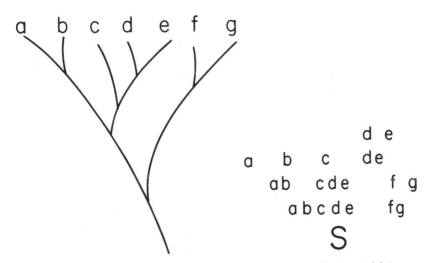

Figure 8.16. A phylogenetic tree (left) expressed as a tree of subsets (right).

nested. In figure 8.16 the subsets have been arranged as a tree to show the natural correspondence with the phylogenetic tree from which they were derived.

More generally, we will call any collection of subsets of S a *tree of subsets of S* if

a) the collection contains S, and

b) any two subsets are either exclusive or nested.

The utility of this concept lies in the fact that to every character-state tree there corresponds a unique tree of subsets, and to every tree of subsets there corresponds a unique character-state tree. Several examples of this bi-unique correspondence are given in figure 8.17. This correspondence makes the addition of any true character-state trees easy; the sum of two true character-state trees is just the union of their trees of subsets, as illustrated in figure 8.18. Notice that with this concept, no direct consideration of speciations on the phylogenetic tree is necessary. To perform the addition of character-state trees using trees of subsets, compare figure 8.18 with figure 8.14. The more complicated addition of figure 8.15 is shown in figure 8.19 using the tree of subsets concept. The naturalness of this biunique correspondence between character-state trees and trees of subsets further justifies the restriction of empty states to those with more than one immediate descendant state.

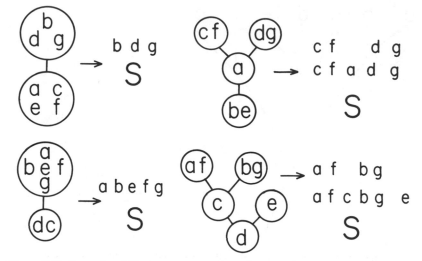

Figure 8.17. Examples of biunique correspondence between character-state trees and trees of subsets.

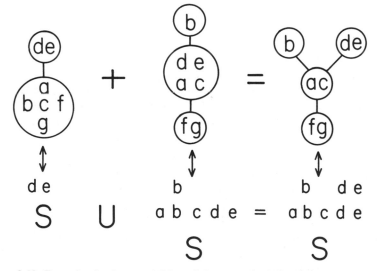

Figure 8.18. Example of union or addition of the trees of subsets derived from the character-state trees shown above them.

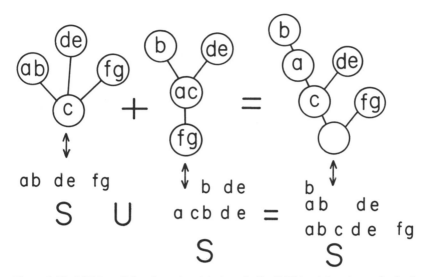

Figure 8.19. Addition of the character-state trees in fig. 8.15 treated as trees of subsets.

The Compatibility of Character-State Trees

In the typical situation, the historically true phylogenetic tree is unknown. Whether one even exists is unknown. But fundamental to the ideas presented here so far is the assumption that a phylogenetic tree exists. This is not a trivial assumption. Subject to this assumption, how to construe character-state trees as hypotheses has been shown above. Although the weight of indirect, circumstantial evidence carefully argued can accrue differential credibility to character-state trees as hypotheses, they are inherently untestable in any direct experimental way, in most cases. Nonetheless, fatal to any argument, no matter how well supported by evidence, are logical contradictions internal to it. Although the absolute truth of any single character-state tree can rarely be checked, it can be determined with complete certainty whether any two character-state trees contradict each other as hypotheses, and groups of logically compatible character-state trees can be identified.

Two character-state trees for S are *compatible* (do not logically contradict each other) if there is some theoretically possible phylogenetic tree for S (not necessarily the historically correct one) such that if it were the correct one, then both character-state trees could (given the appropriate spe-

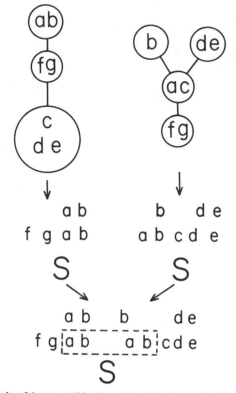

Figure 8.20. Example of incompatible character-state trees, showing the impossibility of adding their trees of subsets.

ciations on this tree) be true. This means that it is logically possible for two compatible character-state trees both to be true at the same time.

The test for the compatibility of two character-state trees is very straightforward: Try to add them. Using their corresponding trees of subsets, this can be done without recourse to a phylogenetic tree (this is why the tree-of-subsets concept is so useful). If the union of their trees of subsets is itself a tree of subsets, then they are compatible; otherwise they are not. Figure 8.20 illustrates the test with a pair of incompatible characters.

Overview

Elsewhere I have reviewed the literature and concepts of character compatibility analysis (Estabrook 1978). More recently, the concept of the tree of subsets used here was introduced by Estabrook and McMorris (1980). A computer program to carry out character compatibility analysis, written by Dr. Kent L. Fiala, Ecology and Evolution Department, SUNY, Stony Brook, New York 11794, is available from him or the author. Among the many recent applications of character compatibility analysis are those in Estabrook and Anderson (1978), Duncan, Phillips, and Wagner (1980), Meacham (1980), and Estabrook (1980). A brief overview of its use is presented here.

1) Character compatibility analysis is an analysis of character-state trees construed as hypotheses of evolutionary relationships to determine whether and how they contradict each other.

2) When the existence and nature of these contradictions is made clear by character compatibility analysis, their resolution based on biological or other considerations is the prerogative of the investigator.

3) In the light of these considerations, revised hypotheses are analyzed as before, and finally (if possible) a collection of mutually compatible characters is put forward as the basis for evolutionary reconstruction.

4) Reconstruction of a phylogenetic tree per se is not possible, of course, but a refined Haase diagram of ancestor-descendant relationships can be constructed by the simple and unambiguous process of adding together all the character-state trees in the collection put forward in step 3.

Literature Cited

Duncan, T., R. B. Phillips, and W. H. Wagner, Jr. 1980. A comparison of branching diagrams derived by various phenetic and cladistic methods. *Syst. Bot.* 5:264–293.

Estabrook, G. F. 1978. Some concepts for the estimation of evolutionary relationships in systematic botany. *Syst. Bot.* 3:146–158.

———. 1980. The compatibility of occurrence patterns of chemicals in plants, In F. A. Bishby, J. G. Vaughan, and C. A. Wright, eds., *Chemosystematics: Principles and Practice*, pp. 379–397. London: Academic Press.

Estabrook, G. F., and W. R. Anderson. 1978. An estimate of phylogenetic relationships within the genus *Crusea* (Rubiaceae) using character compatibility analysis. *Syst. Bot.* 3:179–196.

Estabrook, G. F. and F. R. McMorris. 1980. When is one estimate of evolutionary relationships a refinement of another? *J. Math. Biol.* 10:367–373.

Meacham, C. A. 1980. Phylogeny of the Berberidaceae with an evaluation of classifications. *Syst. Bot.* 5:149–172.

9

Evaluating Characters by Character Compatibility Analysis

CHRISTOPHER A. MEACHAM

The process of estimating evolutionary history begins with the definition of characters. When choosing aspects of organisms for measurement, centuries of precedent may bias the modern systematist. There are a host of reasons why particular features of organisms may be thought uninteresting or uninformative for deriving evolutionary histories. It is not necessary to present arguments here, but only to observe that the impossibility of measuring all features of an organism requires that some features be rejected at the start. Hence, the measurements that are made must reflect some a priori judgments by the systematist, whether based on precedent or on personal biological intuition. Producing an estimate of historical evolutionary relationships among a group of organisms by analyzing their characters can be viewed as a continuation of a process of evaluating characters. Many potential characters are evaluated before character analysis begins.

If a pair of characters is incompatible, then we know with certainty that at least one of the characters involves a parallelism or reversal on the historically true evolutionary tree and is misleading us by making some organisms appear more closely related than they really are. One of the attractive aspects of character compatibility analysis is that, at this basic level, no assumptions are necessary about the mechanism of evolution, i.e., rates of change, etc. We only assume that there is some true tree for the study collection. Character compatibility analysis makes explicit the patterns of agreement and disagreement among characters in a data set. We know that true characters, as defined by Estabrook (this volume), cannot conflict. This fact can be used as a basis for evaluating characters.

Analysis of Compatible Cliques

A set of mutually compatible characters is called a clique (Estabrook, Strauch, and Fiala 1977). Because all true characters must be mutually compatible, all true characters must belong to the same clique. Computer programs are available that can find all the cliques in a data set (see Appendix). Compatibility is a stringent requirement. It is not "easy" for a set of characters to be compatible. The presence of large cliques in a data set tells us that there is considerable agreement among some characters. Our attention should be drawn to the larger cliques, because the pattern of agreement requires explanation. One explanation for the compatibility of characters is logical dependence. If, for example, one character is leaves simple vs. leaves compound and another is leaflets five or fewer vs. leaflets more than five, and we have scored all simple leaves as consisting of a single leaflet, then these characters, by their definition, must be compatible. There is nothing very surprising about finding these characters in the same clique. Other explanations are functional, ecological, or developmental correlations. We would not be surprised to find that floral characters that distinguish pollination syndromes are compatible. If we find, though, a large clique that includes characters whose relationships are not due to obvious correlations, we may feel that the only reasonable explanation left is that these characters are evolutionarily conservative and form an acceptable basis for an estimate of evolutionary history. In this way, analysis of compatible cliques provides a context for applying biological information to the problem of reconstructing the evolutionary relationships among organisms. The reader may refer to Meacham (1980) for further discussion and an example of analysis of compatible cliques.

Probability Analysis of Compatibility Patterns

The idea that compatibility among characters is unlikely to result strictly by chance forms the intuitive basis for much of the justification of clique analysis. As a baseline for evaluating compatibility patterns, it would be useful to know the exact probability of compatibility under a "worst case" model. We can imagine a character that changes so often in the course of evolution that the character states exhibited by the descendants are independent of those possessed by their ancestors. This condition would obtain, for example, if the probability of a character-state transition were greater than 0.5 between branch points on the evolutionary tree (Felsenstein 1980). In this case, knowing the states of the descendants would tell us nothing about those of their ancestors. However, given a set of such rapidly evolving characters, it is still possible for the characters to show compatibilities with each other.

Calculation of Probabilities

Figure 9.1 shows three characters, F_1, F_2, and F_3, for a study collection of eight taxa: a, b, c, d, e, f, g, and h. In the top row is the conventional way of drawing each of these binary characters. In the bottom row is the set diagram that corresponds to each of these characters. For example, character F_1 divides the study collection into two groups: abcd and efgh. In the set diagram, the study collection is represented by the circle and the character F_1 is the curved line that divides the study collection into two subsets, called the character states of F_1. Drawing these binary (two-state) characters as partitions of the study collection makes the possible relationships among characters easy to visualize. The diagrams are oriented so that the state with the smaller number of taxa is on the top. (In the case of F_1, both states are the same size, so the orientation is arbitrary.) These characters exhibit all of the possible relationships among pairs of binary characters. Figure 9.2 shows the relationship of character F_2 with F_3. The smaller state of F_3 is a subset of the smaller state of F_2; F_2 and F_3 are compatible. The corresponding tree is also shown. Figure 9.3 shows the relationship between F_1 and F_3. Here the smaller states are disjoint. Again, F_1 and F_3 are compatible. The third possible relationship is shown by F_1 and F_2 (fig. 9.4). In this case, the smaller states are neither disjoint nor is one a subset of the other. All four possible combinations of character states are present in the study collection. As Wilson (1965) and Le Quesne (1969) recognized, at least one of these characters must show a parallelism or reversal on the true evolutionary tree. The characters are incompatible. In summary, two undirected binary characters are compatible if and only if their smaller states are disjoint or one is a subset of the other. This result is proved in my earlier study (Meacham 1981a).

This result provides a basis for calculating the probability that two characters are compatible at random. What we wish to calculate is, given the number of taxa in the states of a pair of characters, the probability that the two characters are compatible, assuming all distributions of taxa for each character are equally likely. That is, given the size of the states, what is the probability that a pair of characters would be compatible if the taxa were assigned to the states randomly? Figure 9.5 shows diagrams for two binary characters for a study set of 12 taxa. The first character partitions the study

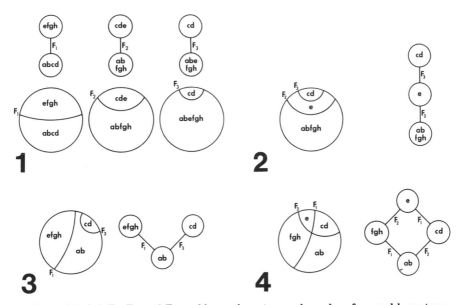

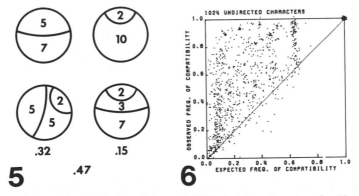

Figures 9.1–9.4. F_1, F_2, and F_3 are binary characters; a, b, c, d, e, f, g, and h are taxa. 9.1. Three character-state trees and the corresponding set diagrams. 9.2. Set diagram of two compatible binary characters with nested smaller states and the corresponding tree. 9.3. Set diagram of two compatible binary characters with disjoint smaller states and the corresponding tree. 9.4. Set diagram of two incompatible binary characters and the corresponding four-cycle.

Figures 9.5–9.6. 9.5. Calculation of the probability that two characters for 12 taxa, one with 2 taxa in the smaller state and the other with 5, are compatible at random. In the top row are the two set diagrams for the two binary characters; in the bottom row are the diagrams of the two ways the two characters can be compatible. 9.6. Composite of scatter plots of expected vs. observed frequencies of compatibility for all 23 data sets listed in table 9.1.

collection into two states that contain 5 and 7 taxa each; the second, into states of 2 and 10 taxa. If we had an urn that contained 5 white balls and 7 black balls, corresponding to the number of taxa in the states of the first character, and drew 2 balls, corresponding to the number of taxa in the smaller state of the second character, then the two characters would be compatible if both balls drawn were the same color. The probability of drawing two white balls is $(5 \times 4)/(12 \times 11) \approx 0.15$. The probability of drawing two black balls is $(7 \times 6)/(12 \times 11) \approx 0.32$. Thus, the total probability that these two characters are compatible is ≈ 0.47. It is clear that this probability depends heavily on the sizes of the smaller states. Two binary characters that divide the study collection in half are much less likely to be compatible at random than two that divide the collection unevenly. This simple model for pairs of undirected binary characters can be extended to sets of undirected binary characters, to directed characters, and to multistate characters (Meacham 1981a).

Present Application

Although the probability model described above could be applied in several ways, the computational complexity of some applications limits their use. One application that requires relatively little computer time is the calculation of the expected number of random compatibilities for each character in the data set. This consists of calculating, for each character, the probability of compatibility with every other character and then summing these values. It is also possible to calculate the standard deviation for the expected number of compatibilities. The observed number of compatibilities can be compared with the number expected to occur at random to assess the extent to which the compatibilities of each character could be explained by random relationships with the other characters in the data set. This is essentially the technique advanced by Le Quesne (1972).

An Analysis of 23 Data Sets

Table 9.1 lists 23 data sets, which together contain 1,024 characters, that were analyzed as described above. These data sets are from a variety of organisms and were collected by many systematists. A sample of this size seems large and varied enough to form the basis for general conclusions. To make comparisons easier, expected and observed compatibilities are di-

Table 9.1. Twenty-three data sets containing 1,024 characters used in analyses

Taxon	Major group	Rank	Author	Characters	EUs
Avena sativa	Plant	Species	Baum and Estabrook (1978)	23	16
Balistoidea	Fish	Superfamily	Matsuura (1979)	32	33
Berberidaceae	Plant	Family	Meacham (1980)	30	15
Beryciformes	Fish	Order	Zehren (1979)	70	26
Oneirodidae	Fish	Family	Pietsch (1974)	30	13
Charadriiformes	Bird	Order	Strauch (1978)	70	227
Cichlidae	Fish	Family	Cichocki (1976)	62	47
Cranchiidae	Mollusk	Family	Voss and Voss (1982)	14	14
Crusea	Plant	Genus	Estabrook and Anderson (1979)	58	17
Fundulus and Profundulus	Fish	Genera	Farris (1969)	35	34
Gymnophiona	Amphibian	Order	Nussbaum (1979)	43	13
Leptopodomorpha	Insect	Suborder(?)	Schuh and Polhemus (1980)	47	8
Lipochaeta diploids	Plant	Genus (part)	Gardner and La Duke (1979)	32	18
Lipochaeta tetraploids	Plant	Genus (part)	Gardner and La Duke (1979)	32	9
Macrouridae	Fish	Family	Okamura (1970)	57	15
Myrceugenia	Plant	Genus	Landrum (1981)	38	45
Orthoptera	Insect	Order	Blackith and Blackith (1968)	92	12
Pygopodidae	Lizard	Family	Kluge (1976)	86	21
Ranunculus hispidus	Plant	Spp. complex	Duncan (1980)	14	87
Tetraodontiformes	Fish	Order	Winterbottom (1974)	66	11
—*	Fish	—	Unpublished	30	30
—*	Fish	—	Unpublished	36	15
—	Fish	—	Unpublished	27	30

*Groups of fish given no formal taxonomic rank.

vided by one less than the number of characters in the data set and so are expressed as frequencies. Scatter plots of expected vs. observed frequency were made by plotting a point at the coordinates $(E_i/n - 1, O_i/n - 1)$ for each character in a data set, where E_i and O_i are the expected and observed numbers of compatibilities for character i, and n is the number of characters in the data set that contains character i. These plots were made with a "cloud" option, which in effect moves each point by a small random displacement from its true position so that points with exactly the same coordinates will not often be plotted on top of each other.

Figure 9.6 is a composite of the scatter plots of all 1,024 characters in the 23 data sets. It is evident from this plot that most characters have more, often many more, compatibilities than would be expected to occur at random. At the same time, a fairly large number of characters fall quite close to the diagonal line and so have close to the number of compatibilities expected at random. At first it might seem that totally random, independent characters would be quite rare, but there are many characters in these data sets that cannot be distinguished from such characters. It is clear that the observed number of compatibilities by itself does not mean much unless compared with the number expected. A character with an observed frequency of compatibility of 0.5 could be explainable by totally random relationships or could be highly nonrandom. Although it could be argued from first principles that random frequencies of compatibility should form a baseline for the observed frequencies, it is probably more convincing to see that of these 1,024 characters, only about six show very many fewer compatibilities than predicted by random probabilities. The probability model appears to describe some basic properties of characters.

Though the composite plot in figure 9.6 shows the general pattern for all the data sets, it obscures the substantial differences among them. Figure 9.7 is the plot for Kluge's data set (1976) for the pygopodid lizards. With the exception of perhaps two characters, the characters of this data set are nonrandom. A contrasting situation is shown by Gardner and La Duke's data set (1979) for the genus *Lipochaeta* (Compositae) (fig. 9.8). Most of the compatibilites in this data set can be explained by random relationships among characters. A similar situation is shown by Duncan's data set (1980) of chromatographic spots for populations in the *Ranunculus hispidus* complex (fig. 9.9). In this data set, the most nonrandom character has 2.2 standard deviations more compatibilities than expected, which is very low when compared with other data sets. Blackith and Blackith's data set (1968) for the Orthoptera is more typical (fig. 9.10). It has a range of characters, from those that are highly nonrandom to those that cannot be distinguished from random ones.

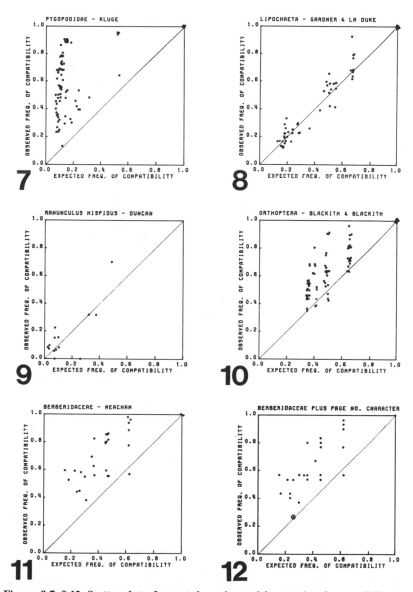

Figures 9.7–9.12. Scatter plots of expected vs. observed frequencies of compatibility for six data sets. 9.7. Kluge's data set (1976) for the pygopodid lizards. 9.8. Composite of Gardner and La Duke's data sets (1979) for the genus *Lipochaeta* (Compositae). 9.9. Duncan's data set (1980) of chromatographic spots from populations in the *Ranunculus hispidus* complex. 9.10. The Blackiths' data set (1968) for the orthopteroid insects. 9.11. Meacham's data set (1980) for the genera of the Berberidaceae. 9.12. Meacham's Berberidaceae data set plus the page-number character from table 9.3 (circled). This plot was done without the "cloud" option.

Probability Analysis of the Berberidaceae Data Set

I have also analyzed the compatibilities in the data set for the Berberidaceae published in Meacham (1980) (fig. 9.11). Table 9.2 lists the characters in descending order of number of standard deviations by which the observed number of compatibilities exceeds the expected number. For the most part, these results support the decisions made about the reliability of characters by analysis of cliques. All but 2 of the 17 most nonrandom characters are among the 20 characters on which the proposed evolutionary history was based. One exception is character 15, tenth on the list. Character 21 is the most nonrandom character in the data set by this measure of probability. This analysis strongly supports the choice, made during secondary clique analysis, of character 21 over 15. The choice of character 27 over 20 is also supported. It is surprising that the second character in table 2, character 3, was rejected during the clique analysis. Of the first 10 characters in the table, character 3 is compatible only with characters 7, 10, 26, and 4. Looking at the distributions of the states of character 3 on the final estimate (Meacham 1980, fig. 18), we find that many of the branches are uniform with respect to this character. Characters that define these branches or parts of them will be compatible with this character. However, characters, like 21 and 30, that determine the basic structure of the estimate suggest very different relationships among the major branches. Thus, even though character 3 is very nonrandom and shows many compatibilities with characters that define local parts of the tree, it is incompatible with characters that indicate relationships among the major groups and do closely agree. If we were to assume that the proposed estimate represents the true evolutionary history, we would call character 3 a conservative but misleading character. The original clique analysis suggests, perhaps marginally, that character 3 is not as reliable as some of the other characters.

A Heuristic Example

The results of these analyses suggest that this probability measure can be used to identify characters that contain little information about the evolutionary history of the study collection. As a heuristic example, I developed a character of the genera of the Berberidaceae that anyone would agree is of no evolutionary significance. Table 9.3 lists these genera with the page number of Willis' *Dictionary*, 1973 edition, on which each appears. Genera whose names appear on even-numbered pages are assigned to one state,

Table 9.2. Probability analysis of the compatibilities in the Berberidaceae data set

Order	Character Number	Expected Compatibilities	Observed Compatibilities	Standard Deviation	Standard Deviation From Expected
1	21	4.6	17.0	1.42	8.72
2	3	5.1	15.0	1.54	6.45
3	25	10.7	24.0	2.23	5.97
4	30	6.7	17.0	1.85	5.59
5	7	13.4	25.0	2.40	4.85
6	10	13.4	25.0	2.40	4.85
7	26	13.4	25.0	2.40	4.85
8	4	7.7	17.0	1.94	4.77
9	2	13.4	24.0	2.40	4.43
10	15	10.6	20.0	2.23	4.18
11	1	18.1	28.0	2.46	4.02
12	8	18.1	28.0	2.46	4.02
13	16	13.4	23.0	2.40	4.01
14	19	13.4	23.0	2.40	4.01
15	29	13.4	23.0	2.40	4.01
16	18	18.1	27.0	2.46	3.61
17	23	8.8	16.0	2.08	3.43
18	28	10.7	18.0	2.23	3.29
19	27	18.1	26.0	2.46	3.21
20	22	7.2	13.0	1.86	3.13
21	9	7.2	13.0	1.89	3.06
22	6	10.7	16.0	2.23	2.39
23	20	18.1	22.0	2.46	1.59
24	13	13.4	17.0	2.40	1.51
25	11	13.4	16.0	2.40	1.10
26	24	13.4	16.0	2.40	1.10
27	5	8.9	11.0	2.08	1.02
28	12	29.0	29.0	0.00	0.00
29	14	29.0	29.0	0.00	0.00
30	17	18.1	17.0	2.46	-0.45

Table 9.3. Page numbers of genera of the Berberidaceae in
the eighth edition of Willis' <u>Dictionary</u> (genera on even–numbered
pages are assigned state A; the others, state B)

Genus	Page number	State
<u>Achlys</u>	10	A
<u>Berberis</u>	133	B
<u>Bongardia</u>	149	B
<u>Caulophyllum</u>	216	A
<u>Diphylleia</u>	370	A
<u>Dysosma</u>	397	B
<u>Epimedium</u>	420	A
<u>Gymnospermium</u>	520	A
<u>Jeffersonia</u>	604	A
<u>Leontice</u>	653	B
<u>Mahonia</u>	705	B
<u>Plagiorhegma</u>	910	A
<u>Podophyllum</u>	925	B
<u>Ranzania</u>	978	A
<u>Vancouveria</u>	1202	A

those on odd-numbered pages to the other. Figure 9.12 is the plot for the
data set including the page-number character. The expected number of
compatibilities of the page-number character is 7.8, with a standard deviation of 1.9. The observed number is 8. This technique clearly identifies
this character, which contains no evolutionary information, as random. It
is interesting to note that the page-number character performs better than
character 17 (number of stamens), which has slightly fewer than the expected number of compatibilities. The compatibilities of character 17 can
be adequately explained by chance. Both the most random character (17)

and the most nonrandom character (21) have 17 compatibilities. The probability model clearly distinguishes the most nonrandom character, which defines three major groups on the final estimate, from the most random character, which was rejected at all stages of the clique analysis, even though both have the same observed number of compatibilities.

Conclusion

Both clique analysis and probability analysis can supply the investigator with useful information about compatibility patterns within a data set. This information, in conjunction with biological information the investigator may possess, can be used to evaluate characters and to support an estimate of evolutionary relationships. In some cases, as shown, this information may demonstrate a great deal of conflict among characters; in other cases it may reveal substantial agreement. This knowledge enables the investigator to judge whether a credible estimate can be advanced from a data set or whether more data are necessary. Whatever method we choose to use, we should require that it tell us when conditions exist under which it is likely to fail.

Acknowledgments

I thank S. G. Poss for making available many fish data sets. This paper is based on Meacham 1981b.

Literature Cited

Baum, B. R. and G. F. Estabrook. 1978. Application of compatibility analysis in numerical cladistics at the infraspecific level. *Can. J. Bot.* 56:1130–1135.

Blackith, R. E. and R. M. Blackith. 1968. A numerical taxonomy of Orthopteroid insects. *Australian J. Zool.* 16:111–131.

Cichocki, F. G. 1976. Cladistic history of cichlid fishes and reproductive strategies of the American genera *Acarichthys, Biotodoma,* and *Geophagus.* Vol. 1. Ph.D. diss., University of Michigan, Ann Arbor.

Duncan, T. 1980. A taxonomic study of the *Ranunculus hispidus* Michaux complex in the Western Hemisphere. *Univ. Calif. Publ. Bot.* 77:1–125.

Estabrook, G. F. and W. R. Anderson. 1979. An estimate of phylogenetic relationships within the genus *Crusea* (Rubiaceae) using character compatibility analysis. *Syst. Bot.* 3:179–196.

Estabrook, G. F., J. G. Strauch, Jr., and K. L. Fiala. 1977. An application of compatibility analysis to the Blackiths' data on orthopteroid insects. *Syst. Zool.* 26:269–276.

Farris, J. S. 1969. The evolutionary relationships between the species of the killifish genera *Fundulus* and *Profundulus* (Teleostei: Cyprinodontidae). Ph.D. diss. University of Michigan, Ann Arbor.

Felsenstein, 1980. A likelihood approach to character weighting and what it tells us about parsimony and compatibility. Talk given at 14th Conference on Numerical Taxonomy, Norman, Okla.

Gardner, R. C. and J. C. La Duke. 1979. Phyletic and cladistic relationships in *Lipochaeta* (Compositae). *Syst. Bot.* 3:197–207.

Kluge, A. G. 1976. Phylogenetic relationships in the lizard family Pygopodidae: An evaluation of theory, methods, and data. Miscellaneous Publications of the Museum of Zoology, University of Michigan, no. 152.

Landrum, L. R. 1981. The phylogeny and geography of *Myrceugenia* (Myrtaceae). *Brittonia* 33:105–129.

Le Quesne, W. J. 1969. A method of selection of characters in numerical taxonomy. *Syst. Zool.* 18:201–205.

———. 1972. Further studies based on the uniquely derived character concept. *Syst. Zool.* 21:281–288.

Matsuura, K. 1979. Phylogeny of the superfamily Balistoidea (Pisces: Tetraodontiformes). *Mem. Faculty Fisheries Hokkaido Univ.* 26:49–169.

Meacham, C. A. 1980. Phylogeny of the Berberidaceae with an evaluation of classifications. *Syst. Bot.* 5:149–172.

———. 1981a. A probability measure for character compatibility. *Math. Biosci.* 57:1–18.

———. 1981b. The estimation of evolutionary history with reference to the Berberidaceae. Ph.D. diss. University of Michigan, Ann Arbor.

Nussbaum, R. A. 1979. The taxonomic status of the caecilian genus *Uraeotyphlus* Peters. Occasional Papers of the Museum of Zoology, University of Michigan, no. 687.

Okamura, O. 1970. Studies on the macrourid fishes of Japan. Morphology, ecology, and phylogeny. Reports of the Usa Marine Biological Station, no. 17.

Pietsch, T. W. 1974. Osteology and relationships of ceratioid anglerfishes of the family Oneirodidae, with a review of the genus *Oneirodes* Lütken. Science Bulletin of the Natural History Museum of Los Angeles County, no. 18.

Schuh, R. T. and J. T. Polhemus. 1980. Analysis of taxonomic congruence among morphological, ecological, and biogeographic data sets for the Leptopodomorpha (Hemiptera). *Syst. Zool.* 29:1–26.

Strauch, J. G., Jr. 1978. The phylogeny of the Charadriiformes (Aves): A new estimate using the method of character compatibility analysis. *Trans. Zool. Soc. London* 34:263–345.

Voss, N. A. and R. S. Voss. 1982. Phylogenetic relationships in the cephalopod family Cranchiidae (Oegopsida). *Malacologia* 23:397–426.

Willis, J. C. 1973. *A Dictionary of the Flowering Plants and Ferns.* H. K. Airy Shaw, ed. 8th ed. Cambridge: Cambridge University Press.

Wilson, E. O. 1965. A consistency test for phylogenies based on contemporaneous species. *Syst. Zool.* 14:214–220.

Winterbottom, R. 1974. The familial phylogeny of the Tetraodontiformes (Acanthopterygii: Pisces) as evidenced by their comparative myology. Smithsonian Contributions to Zoology, no. 155.

Zehren, S. J. 1979. The comparative osteology and phylogeny of the Beryciformes (Pisces: Teleostei). Evolutionary Monographs, no. 1.

Section C

Comparison of Methods

10

The Statistical Approach to Inferring Evolutionary Trees and What It Tells Us About Parsimony and Compatibility

JOSEPH FELSENSTEIN

Cladistics is a term with two distinct meanings. In one, it implies acceptance of a cladist position on classification, the view that all groups in the classification system should be monophyletic. In its other meaning, it signifies an interest in reconstructing phylogenies, without regard to how the classification system is to be set up. This volume is concerned with cladistics in the second sense. The questions of how to construct classifications and how to reconstruct phylogenies are logically separable, so that it would perhaps be better to avoid the word *cladistics* altogether. At present it simply helps perpetuate the confusion of these two tasks.

One would think that the proliferation of alternative methods for reconstructing phylogenies would be a boon, but in numerical "cladistics" it has been more of a plague. The difficulty has been that the proponents of each methodology have been concerned more with developing their own methods and discrediting the alternatives than with exploring the logical interrelations among the methods. I will argue that when we view the matter from the standard scientific framework of statistical inference, the various methods can be seen to depend on the biological assumptions we are making, each having different conditions under which it is to be recommended. We can also see beyond these methods, and identify biologically reasonable circumstances under which there is as yet no adequate method of analysis. For a more complete exposition, the reader may also wish to consult my recent review (Felsenstein 1982).

This work was supported by task agreement DE-AT06-76EV1005 of contract DE-AM06-76RL02225 between the U. S. Department of Energy and the University of Washington.

Consider the following imaginary set of data (table 10.1), collected by scoring six characters in five taxa. Each character has two character states,

Table 10.1. An imaginary set of data

Taxon	Character					
	1	2	3	4	5	6
A	1	1	0	1	1	0
B	1	1	0	0	0	1
C	1	0	0	1	1	0
D	0	0	1	0	0	0
E	0	0	1	1	1	0

0 and 1. If we knew for each character which state was ancestral (= primitive = plesiomorphic) and if in each character this was state 0, then given only characters 1–3 we could apply Hennig's method (1950, 1966) of reconstructing phylogenies. This simply consists of allowing each derived (= advanced = apomorphic) state to define a monophyletic group, so that the state is derived only once on the resulting tree. In the above example, the groups so defined are ABC, AB, and DE. This defines the topology of a phylogeny, shown in Figure 10.1.

One difficulty with Hennigian methods is that we frequently do not know which state is ancestral. Another, and more serious, difficulty is illustrated by the data from characters 4 and 5. These seem to indicate a monophyletic group ACE, which is incompatible (= incongruent) with the monophyletic groups indicated by characters 1–3. Hennig had no recommendation for resolving incompatibilities, beyond restudying the organisms in hope that the incompatibility will prove to be the result of misscored data.

I like to call the resulting impasse "Hennig's Dilemma." A number of methods have arisen that attempt to resolve it. The first is the Camin-Sokal parsimony method. Edwards and Cavalli-Sforza stated a "principle of minimum evolution" (1963) and constructed computer programs to apply it to continuous character data (1964; Cavalli-Sforza and Edwards 1967). For discrete-state data like those considered here, Camin and Sokal (1965) were

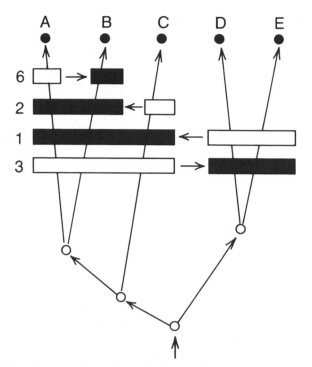

Figure 10.1. The phylogeny obtained from the data in the text using Hennigian methods on characters 1–3. Character 6 is also compatible with this tree. Light bars indicate plesiomorphic states, dark bars apomorphic states.

the first to construct an analogous method, which they referred to as the principle of evolutionary parsimony.

The parsimony or minimum-evolution approach has been the source of several alternative methods, in which different combinations of evolutionary events are allowed to occur and the numbers of different types of inferred character-state changes are to be minimized. These include the Wagner parsimony method (Eck and Dayhoff 1966; Kluge and Farris 1969), the Dollo parsimony method (Le Quesne 1974; Farris 1977), and the polymorphism parsimony method (Farris 1978; Felsenstein 1979). For data with two states 0 and 1, in which state 0 is ancestral, these different methods can be concisely stated as follows:

Camin-Sokal Method: Allow only changes $0 \to 1$. Minimize the number of these changes.

Wagner Method: Allow both forward changes, $0 \to 1$, and backward changes $1 \to 0$. Minimize the total number of these.

Dollo Method: Allow no more than one change $0 \to 1$ in each character. Also allow as many reversions $1 \to 0$ as are needed to explain the data, and minimize the total number of these.

Polymorphism Method: Assume that there can be polymorphism (01) for both states. Assume no more than one change $0 \to 01$ in each character, allow retention $01 \to 01$ of the polymorphism and loss $01 \to 0$ or $01 \to 1$ of either state from the polymorphism. Minimize the number of instances of retention of polymorphism needed to explain the data.

The alternative approach to resolving Hennig's Dilemma is the compatibility method, introduced by Le Quesne (1969) and developed by Estabrook and his coworkers in a long series of papers (e.g., Estabrook, Johnson, and McMorris 1976; Estabrook and McMorris 1980). Compatibility methods find the largest set of characters which are all compatible with the same phylogeny, and give that phylogeny as their result. Compatibility can be viewed as a kind of parsimony (Sankoff, in Le Quesne 1975: 426) in which, instead of counting the changes of state, we count the number of characters which require one or more extra changes of state. By minimizing the resulting quantity over all possible phylogenies, we are in effect maximizing the size of the largest "clique" of mutually compatible characters. Thus, there are two ways of describing the compatibility method: as a method of analyzing conflict between characters or as a method, similar to parsimony methods, for inferring phylogenies.

For all of the above methods there are ordered and unordered variants. Which of these we use depends on whether or not we know which state is ancestral. Figure 10.2 shows a classification of the parsimony and compatibility methods just mentioned. In the left-hand column are the ordered methods, which assume that ancestral states are known (perhaps from outgroup information). Next to each is the corresponding unordered method, in which it is assumed that ancestral states are not known. Two of the unordered methods have been defined in the literature: the Wagner Network method and the unordered compatibility method. The unordered polymorphism method has been mentioned in passing (Felsenstein 1979). The other two methods (unordered Camin-Sokal and unordered Dollo) are new.

It is relatively easy to see how to define these new methods. Let us take the unordered Camin-Sokal method as an example (the unordered Dollo method is defined in an entirely analogous fashion). The method is defined once we know what is the quantity being minimized. Suppose that we are given a tree and want to compute the number of changes required under the unordered Camin-Sokal method. For each character we try to find the reconstruction of the phenotypes at the interior nodes of the tree that re-

ANCESTRAL STATES

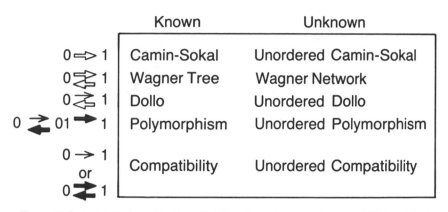

Figure 10.2. A tabular classification of different parsimony and compatibility methods, showing in diagrammatic form the assumptions concerning changes of character state and our knowledge of ancestral states. Small arrows indicate extremely rare events, large unfilled arrows rare events, and large filled arrows common events.

quires the fewest changes of state from ancestral to derived. First we assume that state 0 is ancestral and count the number of changes required under that assumption; then we assume state 1 is ancestral and count the changes. The minimal reconstruction for that character is the lower of these. The number of changes required on the tree is the sum of these for all the characters, as shown in table 10.2.

It is interesting to note that although the unordered Camin-Sokal and Dollo

Table 10.2. Hypothetical example showing computation of number of character states in unordered method

===

	Character						
	1	2	3	4	5	6	7
If state 0 is ancestral	2	1	3	5	4	4	2
If state 1 is ancestral	3	2	2	4	4	5	3
Minimum	2	1	2	4	4	4	2

Sum of minimum values for total number of character-state changes = 19 methods are "unrooted," in the sense that there is no knowledge of the ancestral states, they result in a rooted phylogeny! It might seem that the unordered Camin-Sokal method should be the same as the Wagner Network method, but it is not. In the former case one allows all changes in a character to be $0 \to 1$ or all to be $1 \to 0$, but never allows both $0 \to 1$ and $1 \to 0$ changes to coexist in the same character, as is allowed in the Wagner Network method.

Figure 10.3 shows the results of applying the five ordered methods to the

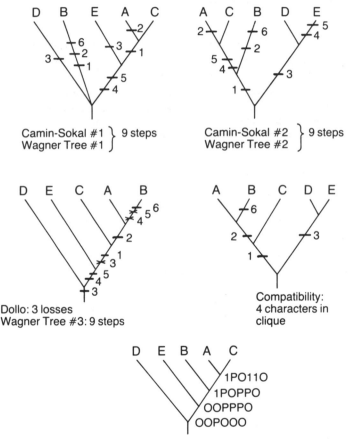

Figure 10.3. Results of applying the ordered parsimony and compatibility methods to the data in the text. Next to each tree are indications of which method yields that tree, how many changes of state are required on it using a Wagner parsimony reconstruction, and the locations of those changes.

sample data set given in table 10.2. The forward changes (bars) and reversions (crosses) and the polymorphic states (P) are shown in each case. In this particular example, some methods give the same trees, or overlapping sets of trees. This should not be accorded undue significance: the pattern of which methods give similar results seems to depend on the exact set of data used, and no general rules are known. Note that under the Wagner Tree method there are ambiguities in the ways that the state changes can be assigned to branches in the tree: there can be many ways, all leading to the same number (9) of total changes. There are multiple assignments for characters 1 and 3 in Wagner Tree # 1, for characters 2, 4, and 5 in Wagner Tree #2, and for character 3 in Wagner Tree #3 in this example. This requires some caution in drawing from a particular assignment of states an inference concerning rates of evolution in different lineages. Readers interested in the unordered methods may wish to try to carry them out by hand on this small data set.

Justification of Different Methods

With so many methods available (and we have not covered pairwise distance methods, such as the Fitch-Margoliash method), attention naturally focuses on how we are to choose among them. What logic will enable us to choose one method over another? The matter is complicated by the presence of two competing logical frameworks: rejectionism and statistical inference.

The difference between these two approaches is illustrated by a simple, if extreme, example. Suppose that we wish to know whether a coin is biased towards heads or towards tails, the possibility of perfect lack of bias being ruled out. We toss the coin five times and obtain three heads and two tails.

The rejectionist believes that "the . . . hypothesis which has been rejected the least number of times is preferred over its alternates" (Wiley 1975: 243), or in accepting the "least refuted hypothesis" (Eldredge and Cracraft 1980: 70). I presume that in the present case this would lead to a preference for the hypothesis that the coin is biased toward heads, since that has been rejected twice (by the two tails) while the other hypothesis has been rejected three times (by the three heads).

The statistical inference approach is the familiar one: we can estimate the probability of heads (in this case 3/5). In standard fashion, we can find the maximum likelihood hypothesis. If we somehow had prior probabilities for the different hypotheses, we could use Bayes' Theorem to compute posterior probabilities for the hypotheses, given the data. We can, in any case, use standard statistical tests to see whether we can reject either hypothesis on the basis of the data. In the present example the hypothesis of a bias toward heads has the higher likelihood, but there is insufficient data to reject a tails bias at the 5 percent significance level (a simple binomial calculation to see if we can reject a bias toward tails gives P = .5).

The statistical approach makes available to us prescriptions (such as the maximum likelihood method) that enable us to go from a probability model, which is assumed to have generated the observed data on an unknown phylogeny, to a choice among the phylogenies. When we do so, the dependence of the resulting method on the biological assumptions is clearly seen. A change of biological assumptions changes the probability model and thereby the method for selecting the best estimate of the phylogeny. The statistical properties of our estimate (such as consistency and efficiency) can also be evaluated in a well-defined way.

For further discussion of some of these issues, the reader may wish to consult the interesting debate between Harper and Platnick (1978). It seems usual to equate the number of times a tree is rejected with the number of extra changes of state the tree requires. This leads to a preference for the most parsimonious tree (for a discussion of parsimony from this perspective, see Kluge, this volume). On the grounds that it makes fewest restrictive assumptions, the Wagner parsimony method has usually been preferred by adherents of the rejectionist approach. It is not at all clear to me why one could not as easily argue that a tree has been rejected once if a character requires even one extra change of state on that tree. If one counted rejections by counting incompatible characters rather than extra changes of state, the rejectionist argument would lead to the choice of a compatibility method. There seems to be an indeterminacy in the choice of a way of counting rejections.

The terms *rejection* and *refutation* create a subsidiary difficulty. If they are meant to be taken literally, why should we ever accept a hypothesis which has been rejected? Is not one rejection sufficient to rule out a phylogeny? Apparently one is not meant to take the terms literally. The number of rejections is to be taken as a measure of goodness of fit, and hypotheses can be accepted even though they have been rejected. But if one is actually talking about goodness of fit, why not be explicitly statistical and avoid a terminology which may be misleading? In any event, it can be seen that there is more common ground between the two approaches than might appear from their terminology. The possibility of reconciliation is real.

Maximum Likelihood

The most general approach to statistical inference is the method of maximum likelihood. It involves computing a likelihood which is the probability of the data, given a particular tree. The likelihood is viewed as a function of the tree, and we seek to find the tree that maximizes the likelihood. It is important to note that the likelihood is *not* the probability of the tree, given the data, but the other way round. We would be interested in knowing the probability of the tree, given the data: it is the a posteriori probability of the tree. But we usually cannot know it, for to compute it we would have to start with a priori probabilities of the various possible trees, and we do not have them. The a posteriori probability is computed by modifying these a priori probabilities in light of the data (see any probability textbook's discussion of Bayes' Theorem). The Bayesian approach to statistics assumes equal a priori probabilities in the absence of any information. This is an arbitrary step and in any event results in the same estimate as maximum likelihood, though the latter avoids the pretense of knowing an a posteriori probability.

The likelihood of a tree is thus not the same thing as the probability of the tree, normal usage in English to the contrary notwithstanding. Because we are computing the likelihood over all possible trees given a particular data outcome, but not over all possible data outcomes, likelihoods do not sum to one, which helps emphasize that they are not to be used as probabilities.

The maximum likelihood method chooses the tree that gives us the highest probability of obtaining the observed data. This procedure is justified, not on abstract philosophical grounds, but because maximum likelihood methods can be proven to have various desirable properties, given certain reasonable assumptions. Chief among these properties is consistency. An estimate is consistent if it is guaranteed to converge to the true value as more and more data are collected. For example, if we are drawing samples from a normal distribution with unknown mean, and are using the sample average to estimate the mean of the underlying distribution, this sample average will converge to the mean as more and more samples are taken, so that the sample average makes a consistent estimate of the mean (in fact, it is also the maximum likelihood estimate).

There may be many different estimates which are all consistent: the sample median will also converge to the mean as samples are taken. We can choose among these on the basis of which estimation method is likely to give estimates closer to the unknown true value. If a method of estimation gives an estimate which varies as little as possible from the true value, it is called *efficient.* Maximum likelihood methods, done correctly, are intrinsically efficient.

Maximum likelihood has three weaknesses. First, in certain cases we find ourselves adding new parameters as we add new data. In such cases the ratio between number of data items and number of parameters does not increase without limit. Frequently this will lead to inconsistency of the resulting estimate of the tree. Second, maximum likelihood estimates, though very efficient for the particular statistical situation in which they have been derived, are sometimes sensitive to small departures from the assumptions. It would be desirable to explore nonparametric statistical methods in hopes of finding estimates of the tree which are robust to such departures from the assumptions. Finally, computation of the likelihood can be difficult. We are attempting to compute the probability of the data, given the tree. This probability is a sum over all the ways that the data could have originated on the given tree. Thus, we must sum the probabilities of all different "scenarios," to use Gareth Nelson's useful term (Hull 1979), which lead to the observed data at the tips of the tree. These involve all the possible assignments of character states to the unobserved interior nodes of the tree, and hence all possible different placements of changes of state on the tree that will yield the observed data. Parsimony methods, in order to evaluate the tree, need only find the one scenario which involves the fewest changes.

There has been some progress in cutting likelihood computations down to size. I have described (Felsenstein 1973b) a method of computing the likelihood which exploits the assumption that different lineages are evolving independently and different characters are evolving independently, and which greatly reduces the computational workload. But the computation still remains a difficult one.

Using the Statistical Inference Approach

There are three ways we can make use of a statistical inference approach in estimating phylogenies. We can make direct use of the method of maximum likelihood to find estimates of evolutionary trees. We can ask under what circumstances the various existing methods are themselves maximum likelihood methods, in order to get some sense of what assumptions might be implicit in those methods. When they do not yield maximum likelihood methods, we can inquire whether they share the desirable statistical properties of that method. There has been work along all three of these lines.

Maximum Likelihood Estimation of Phylogenies

The first use of parsimony methods was made by Edwards and Cavalli-Sforza (1964). They wanted to infer a branching phylogeny of the ancestry of contemporary human populations from blood-group polymorphisms. To do this they assumed (as is reasonable for those loci) that most of the divergence among human populations was due to genetic drift. This provided them with a model from which the probabilities of evolutionary changes of various sorts could be calculated. Both having been students of R. A. Fisher, who invented maximum likelihood, they sought to use that approach but were unable to make it work for purely technical reasons. They adopted a parsimony method, though only as a last resort.

After important progress was made by Gomberg (1966), I found (Felsenstein 1973a) a rapid method of computing the likelihood in this continuous-characters case and produced a computer program to make a maximum likelihood estimate of the phylogeny. To do this it was necessary to ignore a small amount of ancillary information, so that the likelihood I computed was really a restricted maximum likelihood estimate. Such an estimate is the maximum likelihood estimate on the remaining data and so shares many of the properties of maximum likelihood estimates, such as consistency and the ability to compute the variance of the resulting estimate.

Thompson (1975), in a thorough and careful monograph on the problem, took a stricter maximum likelihood approach and produced an iterative computer program. This approach unfortunately brings in more parameters as new loci or new characters are added, so that Thompson's estimate will not generally be a consistent one. More recently, I have made improvements to the iteration strategy of my method (Felsenstein 1981a) which make it computationally practical. At the moment good methods exist for dealing with gene-frequency data, provided one is willing to accept that genetic drift is a reasonable mechanism for the evolution of those frequencies. Use of continuous-character data require information on additive genetic correlations which is not usually available, so that until the appropriate maximum likelihood methods are developed these data can only be treated in a rough fashion (Felsenstein 1973a, 1981a).

DNA-sequence data have also been treated by maximum likelihood methods. The well-known statistician Jerzy Neyman (1971) addressed the problem of inferring phylogenies from DNA sequences, though without producing a practical computational method. Kashyap and Subas (1974) presented a rough approach which was exact only for three species at a time. I have made some inroads (Felsenstein 1981b) into the computational problem of finding the maximum likelihood phylogeny from DNA sequences under a simple model of constant rates of base substitution, but the computation is still so slow as to be on the borderline of feasibility. Kaplan and Langley (1979) have made a maximum likelihood analysis of DNA restriction site data, though their method deals with only two species at a time, estimating divergence times.

Though the situation is likely to change soon, there is currently more protein sequence data available than DNA sequence data. Unfortunately, the same computational problems which make maximum likelihood approaches difficult in the case of DNA sequences make it impossible in the case of protein sequences. Instead of 4 alternative bases, we now have 64 alternative codons or 20 alternative amino acids, and the computation is correspondingly harder. Much remains to be done before we can make efficient use of these data.

Ferris, Portnoy, and Whitt (1979) have applied a maximum likelihood method to the analysis of rates of loss of function of duplicated genes in catostomid fishes. They have used the Likelihood Ratio Test to test constancy of rates of loss. The Likelihood Ratio Test compares likelihoods under different hypotheses. If we have methods of finding maximum likelihood phylogenies under different models of evolution, this technique provides a valuable way of constructing tests of evolutionary hypotheses.

When Are Existing Methods Maximum Likelihood Methods?

By writing down a general formula for the likelihood of a phylogeny under a simple probability model of evolution, it is possible to find conditions under which the maximum likelihood estimate will always be the tree having the fewest changes of character state. This is particularly simple if there are only two character states. In several papers (Felsenstein 1973b, 1979, 1981c) I have looked into this matter. The pattern that emerges is fairly simple.

Suppose that we have a model in which each character has a small probability p of being one which changes rapidly. Otherwise it changes slowly. Suppose that the rate of change of the slowly evolving characters is sufficiently low that, no matter what segments of the phylogeny or what characters are imagined to contain their changes of state, we can order the probabilities of different numbers of changes of state on the tree as follows:

Prob (no steps) $>>$ Prob (one step) $>>$ X $>>$ Prob (two steps) $>>$ . . . $>>$ Prob (N steps) $>>$ . . . $>>$ Y, where $>>$ has its usual meaning of "much greater than." X and Y are explained below. Since a tree having no steps in the character is overwhelmingly more probable than one having even a single step we have clearly assumed that the rate of evolutionary change is very slow in most characters. In the few characters with a high evolutionary rate, the rate is assumed to be so high that these characters effectively contain no information about phylogenetic relationships. (In Felsenstein 1979 I achieved the same result by assuming that these characters were ones which had been badly misinterpreted; the effect is the same.)

There are various possible assumptions we can make about how the fraction of high-rate characters relates to the probabilities of different numbers of changes. The quantities X and Y in the above inequality show two possible sizes for the probability p. The likelihood is the sum of the probabilities of all possible ways of obtaining the observed data, so that it contains terms for both the possibility that a given character is in the high-rate category and the possibility that it is in the low-rate category. If the probability p is of the size Y, then no matter how many steps are required to explain a character on a given tree, the terms reflecting the possibility that the character has a low rate of evolution contribute far more to the likelihood of the phylogeny than do those reflecting the possibility that it is a high-rate character. In this case, it turns out that the likelihood of a tree will reflect exactly the total number of steps which is the minimum required to explain the data on that particular tree. In this case a parsimony method will always choose the same tree as the maximum likelihood method.

At the opposite extreme, p could be of the same size as X. This means that the possibility of a character's being a high-rate character contributes much less to the likelihood than the possibility that it is a low-rate character which requires only one step on the tree, but much more than the possibility that it is a low-rate character which requires two steps. The likelihood of a tree will then reflect the number of characters that can be fitted to the tree with one step required. All characters that require more than one step have most of their contribution to the likelihood coming from the possibility of their being high-rate characters, and the likelihood then does not vary with the number of steps. In this case the maximum likelihood method will maximize the number of characters which fit the tree perfectly, meaning that they require no or only one change of character state. This is precisely what a compatibility method does.

Whether maximum likelihood turns out to be the same as compatibility depends on whether the probability of a character's falling into the high-rate category is greater than the probability of its being a low-rate character which has two steps. If it is less than the probability of its being a low-rate character with a large number of steps, then maximum likelihood is the same as parsimony. Note that both of these conditions require that most characters fall into the low-rate category, so that both assume it is a priori improbable that a typical character will require more than one step.

The reader may have noticed that there is an intermediate zone between the values X and Y. What happens if we assume that the probability of a character's falling into the high-rate category is, say, about the same as the probability of its being a low-rate character which has 4.3 steps? I have treated this recently (Felsenstein 1981c). It leads to an interesting method intermediate between the parsimony and compatibility methods, which I call the "threshold method." In such a case, maximum likelihood will estimate the same tree as the following procedure: to evaluate a tree, compute for each character the minimum number of steps required on that tree. Accumulate a sum by adding up for each character either the number of steps or the number 4.3, whichever is less. Thus, a character requiring 3 steps contributes 3 to the sum, and a character requiring 6 steps contributes 4.3. The result for a hypothetical example is as shown in table 10.3.

If the threshold T (in this case 4.3) is set at less than 2, the threshold method will be exactly the same as a compatibility method, since each term will be either 1 or T, depending on whether or not the character is compatible with the tree. If T is set at a large value, it will never be exceeded, so that the sum will simply be the total number of steps. In between, we will have a method intermediate between parsimony and compatibility. In the data set of table 10.3, the parsimony result is obtained if $T > 2.5$ and the compatibility result if $T < 2.5$. Although no intermediate result is ob-

Table 10.3. Hypothetical example showing evaluation of a tree in a threshold method

	Character						
	1	2	3	4	5	6	7
Changes required	2.0	1.0	4.0	3.0	7.0	2.0	6.0
Threshold	4.3	4.3	4.3	4.3	4.3	4.3	4.3
Minimum	2.0	1.0	4.0	3.0	4.3	2.0	4.3
Sum of minimum values = 20.3							

tainable in that case, on more complex data sets it is usual when T is intermediate to obtain results which are those of neither the parsimony nor the compatibility method. A simple example is given in my paper deriving the threshold method (1981c).

Farris (1969) gave an iterative character-weighting method which has some similarity to the threshold method, though it was not derived using a likelihood argument. For one of his proposed weighting functions it is possible to pass smoothly from a parsimony method through intermediate methods to a compatibility method by changing the value of a parameter. The threshold method is simpler than the ones he suggested and has the advantage that its relation to likelihood arguments is clearer. However, the results of these two approaches might not be much different in practice.

The most pleasing aspect of these results is that they accord well with intuition. We can easily see why a character which changes state very readily will contain little or no phylogenetic information: even close relatives will differ drastically in phenotype. Suppose we find that a tree requires two changes of state in a certain character. If characters with a high evolutionary rate are sufficiently common that two changes of state are most probably a sign of a high-rate character, then once a character requires more than one change of state we certainly should avoid using its information to draw phylogenetic conclusions. Then we should not prefer one tree because a character requires 5 rather than 6 steps. We should base our inferences solely on the number of characters which have zero or one step. This is, of course, part of the logic of the compatibility method.

If, on the other hand, the probability of any character's being a high-rate character is very low, so that all characters are probably low-rate characters, then observing six changes of character state strains our credulity more than observing five changes (though both strain it a great deal). Each additional change of character state which is required makes us more skeptical of the tree, so that we will choose the tree that requires the fewest changes.

The general pattern is quite simple: if a method involves trying to find the tree that minimizes the number of occurrences of some evolutionary event, it implicitly assumes that the event is a priori improbable, so that its occurrence strains our credulity. If there are several possible events and the method minimizes some more than others, then it implicitly assumes that the events it tries hardest to eliminate are the ones which are least probable a priori. Parsimony and compatibility methods make different assumptions about the relative probability of events, specifically the occurrence of extra steps as compared to the assignment of the character to the high-rate category.

The chief advantage of a statistical approach is that it establishes explicitly the connection between biological assumptions, as embodied in the probabilistic model of evolutionary change, and the choice of methods of analysis. The decision can be seen clearly to be a biological decision, based on what the users think they know about the evolutionary processes in operation.

What Are the Statistical Properties of Existing Methods?

The foregoing would be an acceptable logical foundation for the use of parsimony and compatibility methods, were it not for one difficulty. In the justification of either of these methods as a maximum likelihood method, we implicitly assume that changes of character state are rare. If this were true, we should be able to see this in the data: most characters would require no extra steps, and most characters would be members of the same clique. Even if the data had been collected in a way which excluded completely invariant characters, the remaining characters should mostly require only a single change of state.

Unfortunately, few real data sets seem to be this clean. If they were, there would be little controversy over methodology, as most trees could be built by inspection. Most data sets I have seen have a sufficient amount of homoplasy that the maximum likelihood justification for these methods is untenable. Nevertheless, it is possible that these methods, although not maximum likelihood methods, might have acceptable statistical properties, such as consistency. Although maximum likelihood methods are guaranteed to have acceptable statistical properties, other methods may also perform well. An example of this phenomenon is the estimation of the mean of a normal distribution. The maximum likelihood estimate is the sample average, which will be a consistent and efficient estimator of the mean. But the sample median, although not the most efficient estimator, will also be consistent and will be robust with respect to certain departures from normality.

There has been very little work on the statistical properties of parsimony and compatibility methods. An important exception is the work of Cavender (1978, 1981) on the testing of phylogenetic hypotheses. Cavender took the parsimony criterion as given and asked how one could construct a significance test which would be capable of rejecting one tree in favor of another. This is an attempt to correct an important defect of the parsimony method: we have no way of knowing whether a tree requiring (say) two steps more than the most parsimonious tree is to be regarded as significantly worse. There is a corresponding defect in the compatibility method:

are we to regard a clique two characters smaller than the largest clique as significantly worse?

Cavender's work is restricted to a simple two-state model of evolutionary change (similar to the one we have used here) and to consideration of only four species. By considering that unknown true phylogeny which was most likely to yield results favoring an alternative phylogeny, he was able to tabulate the significance points of the number of evolutionary steps. For instance, with 20 characters Cavender (1978) finds that an alternative tree cannot be rejected unless it requires at least nine steps more than the most parsimonious tree.

In a subsequent paper Cavender (1981) shows that his results hold even if the probabilities of change in character states vary from character to character and from one segment of the tree to another, and that they continue to hold when the probabilities of occurrence of states 0 and 1 are unequal. An interesting limitation of this approach, discussed by Cavender, is that the hypothesis test may be inconsistent. A test of a hypothesis is called *consistent* if accumulation of enough data will guarantee rejection of any hypothesis other than the correct one. This is closely related to the notion of consistency of a method of estimation.

In fact, the failure of consistency of the test and of consistency of the estimator are the same phenomenon in this case. The conditions under which Cavender's test will fail to reject a false hypothesis, even though we accumulate vast numbers of characters, are the same as the conditions under which the most parsimonious tree tends to be other than the true tree. Presumably a similar situation exists with respect to the compatibility method, though the matter has not yet been investigated. Whatever its limitations, Cavender's work is a pioneering effort which will form the foundation of much future work.

I have investigated (Felsenstein 1978) the conditions under which parsimony and compatibility will make a consistent estimate of the phylogeny. Given any particular true phylogeny, and assuming that the N characters are a random sample of possible characters, under simple probability models of evolution we can compute the fraction of characters which are expected to be in each of the possible observed configurations. For example, with four taxa and two-state characters, we can compute the probabilities for the 16 possible configurations 0000, . . . , 1111.

If we accumulate enough characters, the observed fractions of them which are in each of these 16 configurations will converge to the expected probabilities. When we reach that situation, what tree is estimated? The details of the argument are presented in my paper (1978), but the essence of the approach is as just described. The only cases I investigated were simple three- and four-taxon cases with two-state characters and particular patterns of rates of evolution in different lineages. This is necessarily a preliminary investigation, but an interesting pattern emerged.

The cases investigated were so simple that parsimony and compatibility methods always would give the same result. It was therefore not possible to investigate conditions under which one would fail but not the other. It was, however, possible to find conditions under which both would fail or both would succeed, in the sense of making a consistent estimate of the topology of the true phylogeny. It turned out that the conditions were that the rates of evolution be either sufficiently small or sufficiently near equal among lineages. When an "evolutionary clock" is in operation (analogous to the "molecular clock"), both methods will be expected to be consistent even if the rates of evolution are not slow. When rates of evolution differ dramatically from one lineage to another, both methods can fail to be consistent even with modest rates of evolution, though for any given disproportionality of evolutionary rates both methods will behave well if evolutionary change is sufficiently improbable a priori.

Failure of these methods to be consistent is fairly disastrous, for it implies that if we expect inconsistency the "least refuted hypothesis" will be more and more likely to be wrong the more data are accumulated! The conditions for consistency are related to the conditions under which the methods are maximum likelihood methods, in that in this case maximum likelihood methods are guaranteed to be consistent. Much work remains, particularly in exploring the conditions favoring the use of parsimony or of compatibility.

Figure 10.4 gives a rough picture of the situation as it now appears, based in part on considerations of consistency and in part on conditions under which the methods are maximum likelihood methods. The region marked "Hennig" is that in which rates of evolution are so low that all characters will be compatible, none requiring any extra steps. In that region Hennig's methods can be applied, both parsimony and compatibility techniques then being identical to each other and to Hennig's methods. If rates of evolution are slightly higher on all characters, a parsimony method will probably perform best, whereas if they are higher on a few characters and remain low on the rest, a compatibility method is to be recommended.

Note that most of the regions of biological interest are marked by a question mark. There is much to be done to develop methods which will not fail when rates of evolution become substantial and unequal, as they may be in many real cases of biological interest. Direct application of maximum likelihood methods may be impractical because of either the computational burden or difficulties in specifying parameters of a realistic model. It

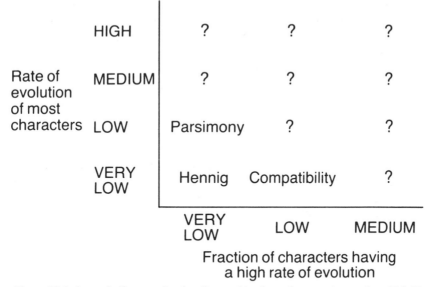

Figure 10.4. A rough diagram showing the combinations of parameters under which Hennigian, parsimony, and compatibility methods can be recommended. A third possible parameter, the inequality of rates in different lineages, has not been shown.

may be more useful to have available robust methods that are not sensitive to small departures from one biological model than to have maximum likelihood methods if these are not robust. In any case, the methods should be evaluated by statistical criteria, rather than defended by the erection of arbitrary philosophical constructs.

Are Program Packages Useful?

A great advantage of a statistical approach is that it makes explicit the connections between the biological assumptions and the choice of a method of analysis. Given that the biological assumptions concerning the relative frequencies of different kinds of changes will differ from study to study (indeed, from character to character), it is desirable to have available a variety of methods of analysis. I have been distributing PHYLIP, a package for inferring phylogenies, written in Pascal and designed to work on almost all computers which have Pascal compilers (which is everything from IBM 370s to microcomputers; see Appendix 10.1 at the end of this paper).

One criticism of the availability of packages is that they may encourage the random running of programs without careful consideration of the biological reasons for selecting one method. It is certainly a temptation for the systematist to run all available programs, whether their assumptions make biological sense in the particular case or not, and then choose the result which is closest to the investigator's preconceptions. In such a case it should be realized that the procedure is circular, and that it is the preconceptions, not the programs, which are yielding the result.

It seems pointless to abandon program packages: the only alternatives are to have all systematists write individual customized programs or to imagine that there is one all-purpose method which can be recommended in all cases without reference to the biology. The former would be desirable, but seems impractical in view of the discomfort induced in most systematists by the proximity of a computer. The latter is a misuse of numerical and computer methods.

Is Statistics Useful?

I am not sanguine about the rapid spread and growth of statistical approaches to the reconstruction of phylogenies and the testing of evolutionary hypotheses. Statisticians tend to put their efforts into fields where the money is, such as medical clinical trials, and systematists are uncomfortable with numerical approaches, preferring methods which seem to promise certainty to those which explicitly acknowledge uncertainty.

The most frequent criticism of the statistical approach is that it is based on unrealistically oversimplified models of evolutionary change. Certainly no biologists believe that the characters they study are changing randomly and independently, at a constant rate, between two symmetrical states. Of what use, then, are conclusions based on these assumptions?

Of considerable use. If a method fails with such a simple model, we have no particular reason to believe it will behave itself when analyzing data produced by real evolutionary processes. We can thus use the oversimplified models to investigate the properties of proposed methods and to rule out candidates for use on real data. We have also seen that we can use the tractability of simple models to get some sense of what kinds of features of evolution will cause trouble.

The difficulty we face is that we know too little to specify a realistic model of evolutionary change. Even if we could do so, it would not be mathematically tractable. In this sense the advocacy of total realism is a counsel of despair. The best that we can do is to use the most reasonable model we can find which is also tractable. To the statistical errors of measurement and estimation we calculate under such a model, we must then add an extra dose of uncertainty, corresponding to our uncertainty about the biological processes.

This may seem to place statistical approaches at a disadvantage, but I am convinced that the same problem arises in all other fields to which statistics is applied. We are never confident that the coin we toss behaves identically and independently on each toss, and this necessarily makes our estimate of the statistical uncertainty too small. The situation in systematics differs from this only in the size of this extra uncertainty, and we have at least the hope that advances in our understanding of evolutionary processes can reduce that uncertainty.

It is not clear whether there is any acceptable alternative. Other frameworks either reduce to a statistical one, while using a terminology which seems to be different, or promise the fools' gold of absolute certainty. One must hope that certainty in systematics is not attainable, for if it were, it would remove systematics altogether from the realm of science.

Acknowledgments

I am indebted to Walter Fitch for constructive criticisms of an earlier draft of this paper.

Appendix 10.1 Availability of PHYLIP

PHYLIP, a package for inferring phylogenies, is available from me free of charge. Interested readers should send a magnetic tape (a small one will do), which will be returned with the package and its documentation written on it in standard ANSI format. The package includes programs to perform the Camin-Sokal, Wagner, Dollo, and polymorphism parsimony methods, a compatibility method, the Fitch-Margoliash and least-squares methods for distance data, a parsimony and a compatibility method for DNA sequences, and two maximum likelihood methods, one for gene frequency data and one for DNA sequences. There is also a program to generate all monothetic subsets. All programs are written in a highly standard subset of Pascal and will work on nearly every Pascal compiler available. A CP/M-80 version on 8-inch diskettes is available from Thomas K. Wilson, Department of Botany, Miami University, Oxford, OH 45056.

Literature Cited

Camin, J. H. and R. R. Sokal. 1965. A method for deducing branching sequences in phylogeny. *Evolution* 19:311–326.

Cavalli-Sforza, L. L. and A. W. F. Edwards. 1967. Phylogenetic analysis: models and estimation procedures. *Evolution* 32:550–570. (Also published in *Amer. J. Human Genet.* 19:233–257.)

Cavender, J. A. 1978. Taxonomy with confidence. *Math. Biosci.* 40:271–280.

———. 1981. Tests of phylogenetic hypotheses under generalized models. *Math. Biosci.* 54:217–229.

Eck, R. V. and M. O. Dayhoff. 1966. *Atlas of Protein Sequence and Structure 1966.* Silver Spring, Md.: National Biomedical Research Foundation.

Edwards, A. W. F. and L. L. Cavalli-Sforza. 1963. The reconstruction of evolution. *Ann. Human Genet.* 27:105. (Also published in *Heredity* 18:553.)

———. 1964. Reconstruction of evolutionary trees. In V. H. Heywood and J. McNeill, eds., *Phenetic and Phylogenetic Classification*, pp. 67–76. Systematics Association Publication No. 6.

Eldredge, N. and J. Cracraft. 1980. *Phylogenetic Patterns and the Evolutionary Process.* New York: Columbia University Press.

Estabrook, G. F., C. S. Johnson, Jr., and F. R. McMorris. 1976. A mathematical foundation for the analysis of cladistic character compatibility. *Math. Biosci.* 29:181–187.

Estabrook, G. F. and F. R. McMorris. 1980. When is one estimate of evolutionary relationships a refinement of another? *J. Math. Biol.* 10:367–373.

Farris, J. S. 1969. A successive approximations approach to character weighting. *Syst. Zool.* 18:374–385.

———. 1977. Phylogenetic analysis under Dollo's Law. *Syst. Zool.* 26:77–88.

———. 1978. Inferring phylogenetic trees from chromosome inversion data. *Syst. Zool.* 27:275–284.

Felsenstein, J. 1973a. Maximum-likelihood estimation of evolutionary trees from continuous characters. *Amer. J. Human Genetics* 25:471–492.

———. 1973b. Maximum likelihood and minimum-steps methods for estimating evolutionary trees from data on discrete characters. *Syst. Zool.* 22:240–249.

———. 1978. Cases in which parsimony or compatibility methods will be positively misleading. *Syst. Zool.* 27:401–410.

———. 1979. Alternative methods of phylogenetic inference and their interrelationship. *Syst. Zool.* 28:49–62.

———. 1981a. Evolutionary trees from gene frequencies and quantitative characters: finding maximum likelihood estimates. *Evolution* 35:1229–1242.

———. 1981b. Evolutionary trees from DNA sequences: A maximum likelihood approach. *J. Mol. Evol.* 17:368–376.

———. 1981c. A likelihood approach to character weighting and what it tells us about parsimony and compatibility. *Biol. J. Linn. Soc.* 16:183–196.

———. 1982. Numerical methods for inferring evolutionary trees. *Quart. Rev. Biol.* 57:379–404.

Ferris, S. D., S. L. Portnoy, and G. S. Whitt. 1979. The roles of speciation and divergence time in the loss of duplicate gene expression. *Theor. Pop. Biol.* 15:114–139.

Gomberg, D. 1966. "Bayesian" post-diction in an evolution process. University of Pavia, Italy. Manuscript.

Harper, C. W., Jr., and N. I. Platnick. 1978. Phylogenetic and cladistic hypotheses: A debate. *Syst. Zool.* 27:354–362.

Hennig, W. 1966. *Phylogenetic Systematics.* D. D. Davis and R. Zangerl, trans. Rpt. 1979. Urbana: University of Illinois Press. Originally published as *Grundzüge einer Theorie der phylogenetischen Systematik.* Berlin: Deutscher Zentralverlag, 1950.

Hull, D. L. 1979. The limits of cladism. *Syst. Zool.* 28:416–440.

Kaplan, N. and C. H. Langley. 1979. A new estimate of sequence divergence of mitochondrial DNA using restriction endonuclease mappings. *J. Mol. Evol.* 13:295–304.

Kashyap, R. L. and S. Subas. 1974. Statistical estimation of parameters in a phylogenetic tree using a dynamic model of the substitution process. *J. Theor. Biol.* 47:75–101.

Kluge, A. G. and J. S. Farris. 1969. Quantitative phyletics and the evolution of anurans. *Syst. Zool.* 18:1–32.

Le Quesne, W. J. 1969. A method of selection of characters in numerical taxonomy. *Syst. Zool.* 18:201–205.

——. 1974. The uniquely evolved character concept and its cladistic application. *Syst. Zool.* 23:513–517.

——. 1975. Discussion of preceding presentations. In G. F. Estabrook, ed., *Proceedings of the Eighth International Conference on Numerical Taxonomy,* pp. 416–429. San Francisco: W. H. Freeman.

Neyman, J. 1971. Molecular studies of evolution: A source of novel statistical problems. In S. S. Gupta and J. Yackel, eds., *Statistical Decision Theory and Related Topics,* pp. 1–27. New York: Academic Press.

Thompson, E. A. 1975. *Human Evolutionary Trees.* Cambridge: Cambridge University Press.

Wiley, E. O. 1975. Karl R. Popper, systematics, and classification: A reply to Walter Bock and other evolutionary taxonomists. *Syst. Zool.* 24:233–243.

11

Application of Compatibility and Parsimony Methods at the Infraspecific, Specific, and Generic Levels in Poaceae

BERNARD R. BAUM

The purpose of this paper is to present ideas from experiences I have had with numerical parsimony and compatibility methods which have been applied to taxonomic problems of different kinds, and to comment on these from the user's point of view. Part of the results of these methods is reported here for the first time; this will be dealt with in greater detail. For this study, however, all the original and new analyses have been run with the IBM 3033 computer currently in use in the Department of Agriculture, Ottawa.

This paper summarizes my experience with character compatibility analysis, developed by Estabrook and associates (Estabrook, Johnson, and McMorris 1976a, 1976b; Estabrook and Meacham 1979); and with parsimony methods developed by Camin and Sokal (1965) and by Farris, i.e., the computation of Wagner networks (1970) and the Wagner distance procedure (1972a). Specifically, the programs used were CLADON 1, CLADON 2, and CLADON 3 (Bartcher 1966); COMPATIBILITY (Estabrook and Fiala 1976); CLINCH 1 (Fiala and Estabrook 1977); CLINCH 3 (Fiala 1980); CLAD/OS (Farris 1972b) and WAGNER 78 (Farris 1978).

At the infraspecific level the evolutionary units (EUs) were *Avena* cultivars; at the specific level they were *Avena* species; and at the generic level they were genera of the tribe Triticeae.

Numerical Cladistic Methods in General

All the numerical cladistic methods so far developed are based on the assumption that cladogenesis is divergent, i.e., there is no accommodation for anastomosing branching. Sneath (1975) explored problems of and approaches to cladistic representation of reticulate evolution. Reticulate evolution is common at the infraspecific level. An example from cultivated plants, where reticulation is induced and then channelled by human beings, and its bearing on classificatory problems is presented by Baum and Lefkovitch (1973).

Reticulate evolution at the species level in vascular plants is a common phenomenon. Probably a third or more of all plants are polyploids (Stebbins 1966; Grant 1971). The cytogenetic method has enabled investigators to elucidate many amphidiploids and unravel whole syngameons (Grant 1971).

At the generic level and above, reticulate evolution has not yet been found by cytogenetic analyses, with a few exceptions, such as in some Triticeae. This is understandable, since at the generic level, and more so at higher levels, intertaxon experimental crosses are not possible.

It thus appears that, at least for vascular plants, the numerical cladistic methods that have so far been devised are inappropriate for study collections at the infraspecific level. At the species level they can more profitably be used, especially when the study collection has already been investigated cytogenetically. Hence, when cladistic analyses are confined separately to species of the same ploidy level or to those with the same genome, they can be very useful for unraveling relationships for which conventional cytological methods are sometimes limited (Baum 1975).

At the generic level and above, the methods of numerical cladistic analysis are extremely valuable and useful. They are perhaps the only means of generating cladistic hypotheses with various kinds of data. As mentioned above, cytogenetic methods have intrinsic limitations at those levels. More promising are molecular data, e.g., nucleotide sequences (Gingras and Roberts 1980).

Infraspecific Level

The rationale for conducting a cladistic analysis at the infraspecific level was to compare the results of compatibility analysis with a known, dichotomously branching cladogram. The known cladogram was obtained from pedigree data generated using the computer method of Baum and Thompson (1970). The 16 carefully selected cultivars have the pedigree shown in figure 11.1.

Character compatibility analysis of the data revealed one largest clique of 7 among 23 characters. The characters were taken from Stanton (1955), but definition of states, their arrangement into transformation series, and coding was done by us (Baum and Estabrook 1978). After some characters were altered so that their state trees were arranged in reverse order, another compatibility analysis revealed one clique with the greatest number of characters, 6. The tree generated from the first-mentioned clique is shown in figure 11.2 and that of the last-mentioned clique appears in figure 11.3.

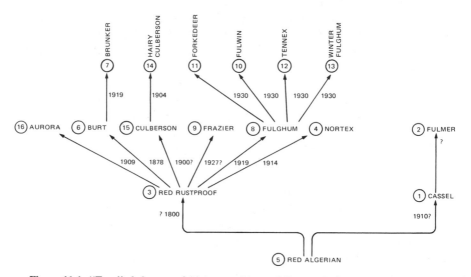

Figure 11.1. "True" cladogram of 16 *Avena* cultivars. All were obtained by breeding through selections; thus, there is no reticulate evolution in this study collection.

Clearly both trees are far from resembling the true cladogram. What went wrong? Why do so few characters correctly reflect the historical relationships among the 16 cultivars?

The first possibility is that the "true" cladogram is not true. Although it is a possibility, the chances are very slim, because the EUs were carefully selected from among the best-documented cultivars just to minimize that possibility.

The second possibility is that the characters used cannot reflect the true evolutionary situation. In other words, if all possible characters were available, there could be a clique with characters that would be true to the ev-

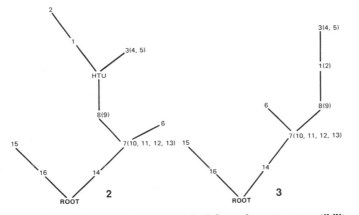

Figures 11.2, 11.3. Primary cladograms obtained from character compatibility analyses. HTU denotes Hypothetical Taxonomic Unit. See text.

olutionary situation. Even though two cliques of compatible characters were found, the two cliques are irrelevant to the true cladogram. This raises a very serious problem, at least for the investigator who uses morphological characters. Thus, the organophyly of the character-state trees may not necessarily reflect phylogenesis.

Third, at the infraspecific level interbreeding between EUs (the cultivars, in this case) occurs as a rule because of lack of reproductive barriers. This, however, is minimized to the point of nonoccurrence in this case, because human selection and maintenance of purity are very drastic and ensure that the cultivars are kept in isolation. Furthermore, these plants are predominantly self-fertilized. A fourth possible explanation is that at the infraspecific level, and especially at the cultivar level, evolution of characters proceeds at a fast rate.

A colleague, Dr. D. B. O. Saville (pers. comm., 1979), drew my attention to another possibility: that ecological selection, both random and human-directed, in response to one or more environmental elements can cause convergent similarities in certain characters. If this is true, then what is the mechanism that causes such convergences?

This investigation at the infraspecific level has actually generated more questions that it has answered. The basic question is this: Do all characters potentially reflect the cladogram? If not, what are the rules that determine whether they do so in the course of evolution? For the user of compatibility analysis the same question will always exist, especially when morphological characters are used: If a clique of compatible characters is found, what is the relevance of that clique?

Specific Level

The purpose of the investigation of the cladistic relationship among *Avena* species was to unravel, using morphological characters, the possible evolutionary relationships among EUs within monophyletic groups. The monophyletic groups had previously been identified mainly by cytogenetic methods. Polyploidy is common in *Avena*. Four genomes were found in *Avena*, as were a number of polyploid groups with particular genome combinations (Rajhathy and Thomas 1974). Cytogenetic methods have intrinsic limitations for resolving the relationships among species that belong to the same genome group. Thus, using a total of 28 morphological characters, a cladistic analysis of the 14 diploids (22 characters) was carried out separately from another cladistic analysis of the 8 hexaploids (19 characters). There was no need for such analysis at the tetraploid level, because each of the different groups is too small, having only two members each.

At the diploid level the A-genome and C-genome species (Baum 1977, Rajhathy and Thomas 1979) were incorporated together in the same analysis. In both Camin and Sokal's and Farris' Wagner procedures, the C-genome species were monophyletic on the generated trees (figures 11.4–11.5). In figure 11.5, top left, the C-genome species form a paraphyletic group. The A-genome species in most of the cladograms of figure 11.5 are split into two groups, but this bears no relevance to the finding that the C-genome species form a monophyletic group. The fact that the A-genome species tend to form two monophyletic groups is suggestive of genetic (as opposed to cytogenetic) evolutionary divergence in this genome group. Character compatibility analysis yielded various primary trees. When the character suite did not include the character representing genome information, compatibility analysis yielded eight primary trees, of which four (figure 11.6, upper row) are acceptable because the EUs with the C-genome form a monophyletic group. The other four trees (figure 11.6, bottom row) are not acceptable hypotheses. With the genome character included, compatibility analysis yielded five primary trees (figure 11.7), all of which are acceptable hypotheses.

At the hexaploid level only Wagner networks were generated. Since one EU, *Avena sativa* "Fatuoid," is known from genetic and cytologic evidence to be a mutant of *A. sativa*, it is expected to be a sister EU on the cladogram. In the networks obtained (shown in figure 11.8), this expectation never materialized. Otherwise, the cladograms are acceptable.

From the user's point of view, what have we learned so far? First, knowledge of cytology and preferably of the genomic groups would be extremely

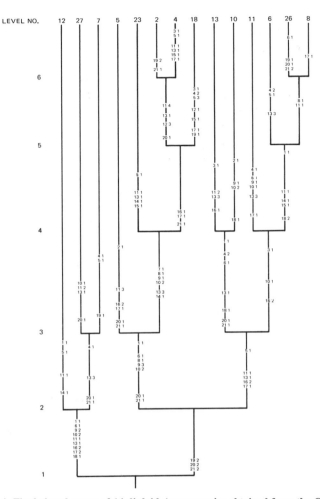

Figure 11.4. Final clantdogram of 14 diploid *Avena* species obtained from the Camin-Sokal procedure. Note that at level 5 the 3 C-genome species cluster together. See table 11.1 for EU identification and text for details.

useful before any attempts to carry out cladistic analysis based on morphological characters. Second, we still do not fully understand the relationship between genetic change and its effect on the phenotype. This is important because in plants, and especially in grasses, the number of morphological characters that are species specific is relatively low. Thus, the determination of characters and the definition of character-state trees

Table 11.1. Data matrix for diploid species of <u>Avena</u>

EU	Genome Group	Characters and States
2 HISPANIC 1	A	1 0 0 0 0 0 0 0 0 4 0 3 1 0 1 1 0 4 4 5 1
4 BREVIS 2	A	1 0 1 0 1 0 0 0 0 0 5 0 4 1 1 1 0 0 2 4 4 1
5 CANARIEN 3	A	1 1 0 0 0 0 1 1 1 2 3 0 0 0 0 0 1 0 2 4 4 1
6 CLAUDA 4	C	1 0 1 2 2 1 1 2 4 3 2 0 3 1 1 1 1 2 2 2 2 0
7 DAMASCEN 5	A	1 0 0 2 1 0 1 2 2 2 1 0 3 0 0 0 0 1 1 1 1 1
8 ERIANTHA 6	C	1 0 1 0 1 1 1 4 3 3 0 1 1 1 1 0 2 2 2 2 0
10 MATRITEN 7	A	1 1 0 2 1 0 1 2 3 2 1 0 2 0 0 0 1 2 2 3 3 1
11 LUSITANI 8	A	0 0 1 1 0 1 2 3 2 1 0 3 0 0 1 0 0 2 2 2 1
12 HIRTULA 9	A	2 0 0 0 1 0 1 2 2 2 2 0 1 0 0 0 0 2 0 0 0 1
13 LONGIGLU 10	A	1 0 1 2 1 0 1 2 4 4 3 0 6 0 0 3 1 1 2 3 3 1
18 NUDA 11	A	1 0 1 2 3 0 0 0 0 0 0 1 5 1 1 4 0 0 3 3 4 1
23 STRIGOSA 12	A	1 0 0 0 1 0 0 0 0 0 1 0 3 2 1 2 2 0 2 3 3 1
26 VENTRICO 13	C	1 0 1 0 1 0 1 1 4 3 3 0 1 1 1 1 1 2 3 3 4 0
27 WIESTII 14	A	1 0 0 1 0 0 1 2 2 1 3 0 6 0 0 0 1 0 2 1 1

Note: Numbers left of the EUs appear in fig. 11.4; numbers right of the EUs appear in figs. 11.6 and 11.7.

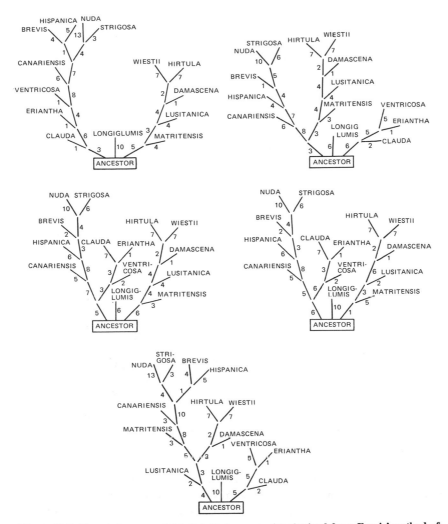

Figure 11.5. Five cladograms of 14 diploid *Avena* species obtained from Farris' method of computing Wagner trees. Steps are indicated by numbers on internodes.

will determine to a great degree the results of cladistic analysis. The important point is that there is no check or test for validity of these determinations and definitions. Such a check for validity would require knowledge of the relationship between genetic change and its effect on the phenotype. It thus appears that this could be a weakness of morphological taxonomy, as contrasted to, say, nucleotide sequence data. However, it is not ger-

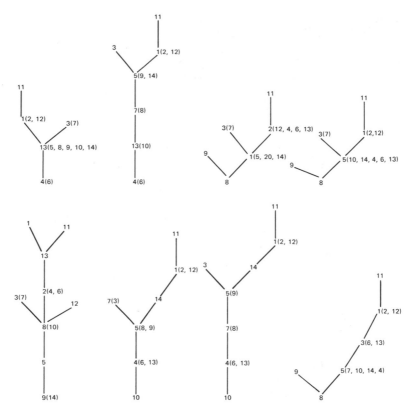

Figure 11.6. Eight primary cladograms of 14 diploid *Avena* species obtained from character compatibility analysis without the character on genome information. See text.

mane to assess the merit of morphological taxonomy here. Third, it follows from the above that it might be promising to use characters of molecular nature, e.g., protein sequences or DNA sequences, and morphological characters in combination. This might provide closer approximations to the true cladogram and give us estimates of degree of compatibility between morphological characters and molecular ones in each case. These might then give us a clue about how the phenotype, expressed in morphological characters, is related to the genotype in terms of molecular characters.

In most cases, at present, morphologists do not use molecular characters and molecular biologists do not use morphology. Thus, from the practical point of view morphologists ought to be aware of the limitation of the char-

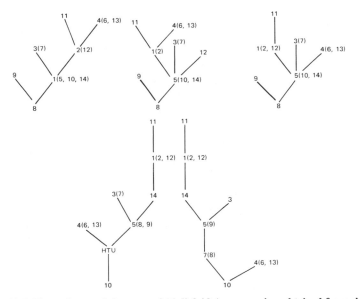

Figure 11.7. Five primary cladograms of 14 diploid *Avena* species, obtained from character compatibility analysis with the character on genome information. See text.

acters they employ. Moreover, it might be useful if morphologists would introduce external criteria, such as genome information or conspecificity of two EUs, as a check for confidence in the results. This opinion on morphology does not imply that morphological taxonomists are relieved from their responsibilities for thoroughly studying characters.

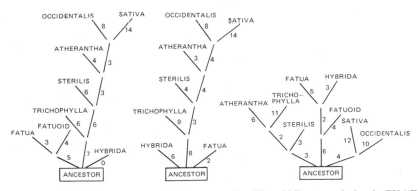

Figure 11.8. Cladograms of hexaploid *Avena* species. The middle one excludes the EU "Fatuoid." See text.

Generic Level

The reason for conducting a cladistic analysis of Triticeae based on morphological characters is to obtain estimates of possible evolutionary relationships in the tribe as a basis for future evolutionary studies. More is known about Triticeae than about any comparable tribe of vascular plants, but some of this knowledge is uneven and some controversial. Because of this, the EUs are still not satisfactorily defined morphologically. Even though there is evidence, primarily from cytology, of some reticulate evolution in Triticeae, the three methods of numerical cladistic analysis used, yielded results that might increase our insight into this tribe in spite of their suspected inappropriateness in this regard.

Twenty-eight EUs were defined for this study, and in some runs a 29th EU, "ROOT," was added. The number of characters scored was 35, but because of inapplicability due to lack of comparison in a few EUs, only 32 characters were used.

Character compatibility analysis was performed on the data matrix, with character-state trees hypothesized as in figure 11.9. The states of the characters not in figure 11.9 are ordered in linear fashion: A, B, C. . . . The characters of ROOT consisted of a vector containing all the plesiomorphic states. For undirected analyses the character-state trees were obviously regarded as polarized but remained ordered, and ROOT was not included in the computer runs. Compatibility analysis, using both directed and undirected options, determined all the primary characters that were mutually compatible. The largest cliques were composed of four characters. The undirected option depicted the same cliques as the directed option and two additional cliques. Of those two, the primary network obtained from one is entirely different from all the other networks. This clique was not considered further because it did not meet the external criterion that EUs *Hordeum* and *Critesion* must be sister groups for the tree to be admissible. One clique yielded a tree (figure 11.10) that is very close to that of another (figure 11.11)—not surprisingly, because three characters are common to both. Four cliques have three characters in common, so that the trees obtained from them have a high resemblance.

To obtain the complete trees, compatibility analyses of subcollections of unresolved EUs on the primary trees were carried out according to the procedure in Estabrook, Strauch, and Fiala (1977). The resulting trees were then added to the primary tree in each case in place of the unresolved areas. Since unresolved areas on the secondary trees still remained, the same procedure was repeated. Subsequent compatibility analyses were needed for tertiary characters and only in some cases for quaternary characters also.

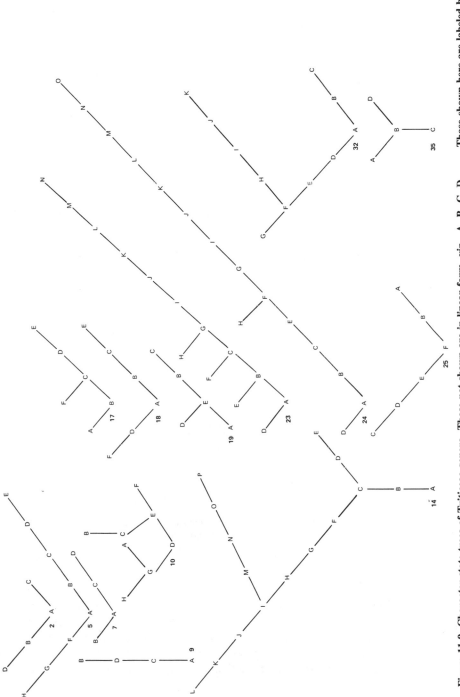
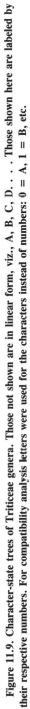

Figure 11.9. Character-state trees of Triticeae genera. Those not shown are in linear form, viz., A, B, C, D . . . Those shown here are labeled by their respective numbers. For compatibility analysis letters were used for the characters instead of numbers: 0 = A, 1 = B, etc.

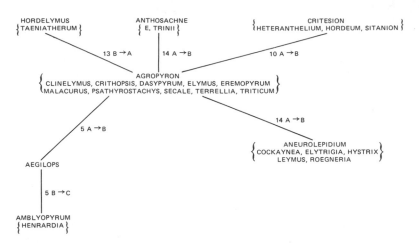

Figure 11.10. Primary network of Triticeae genera obtained from clique 5, determined from character compatibility analysis of undirected characters. See text and figure 11.9.

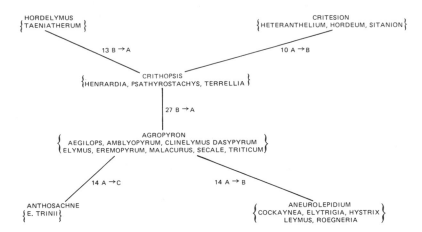

Figure 11.11. Primary network of Triticeae genera obtained from clique 2, determined from character compatibility analysis of both directed and undirected characters. See text and figure 11.9.

In total, 66 cliques were used to complete different parts of the trees. These include the largest 4 cliques for the primary trees, 18 largest cliques for the secondary trees, 40 largest ones for the tertiary trees, and only 4 to resolve the quaternary trees. All this resulted in a number of alternative hypotheses of cladistic relationships among the genera of Triticeae (figures 11.12–11.19).

The Triticeae data matrix mentioned above was also subjected to the

Wagner method. Here too the same external criterion for admissibility of cladograms was used, viz., that *Hordeum* and *Critesion* be sister EUs on the resulting cladograms. For this reason I did not seek the tree or trees with fewest steps; but I sought instead those with fewest steps that also met the external criterion. For this the Wagner computations were performed 29 times, each time with another EU as ancestor (ancestor for the run—this does not have any bearing on the ancestor or root of the tree). The 29th time the ancestor was ROOT, which was made up of the vector of all the plesiomorphic states that were assumed in the hypothesized transformation series just prior to and for compatibility analysis. This resulted in four alternative networks (figures 11.20–11.23), one of which (figure 11.23) turned out to be the tree with ROOT as ancestor.

The Triticeae were also analyzed by means of the Wagner distance procedure. For this analysis data on crosses were amalgamated with the morphological data in the following manner: a dissimilarity matrix was computed from the morphological data; every element of the dissimilarity matrix was squared; the matrix of products was added to the matrix of crosses—a 0, 1 matrix; and every element of the new matrix was divided by 2 and its square root calculated (Baum 1978). This new distance matrix was then subjected to the Wagner distance procedure. The network obtained (figure 11.24) is different from the various networks and trees previously obtained by different procedures from different data.

Although the external criterion of admissibility, as defined, is not met in this network, it is still an acceptable one. This is because the patristic distance between *Hordeum* and *Critesion* is the shortest one relative to other EUs. It is even shorter, viz. 53, than the distance between the two sister genera *Hordelymus* and *Critesion*, viz. 59.

How can we assess the differences between the various results? How do we make a choice between the various cladistic hypotheses? Should we make a decision at this stage of the study for classificatory purposes?

We could unequivocally determine the total length of the various trees and use this as a difference between them and choose the shortest tree. Of the newly developed criteria in the case of compatibility analysis, we could compute the probability that a clique occurs at random (Meacham, this volume) and subsequently choose the clique that is least likely to occur at random but has occurred, and repeat this for all the cliques generated from the secondary trees, and so forth. But this is not enough, because the topology of the tree is also important. To compare topologies, methods of comparing classifications (Rohlf 1974) and consensus methods might be useful. Intuitive assessments primarily based on the recognition of monophyletic groupings and their biological significance might certainly be use-

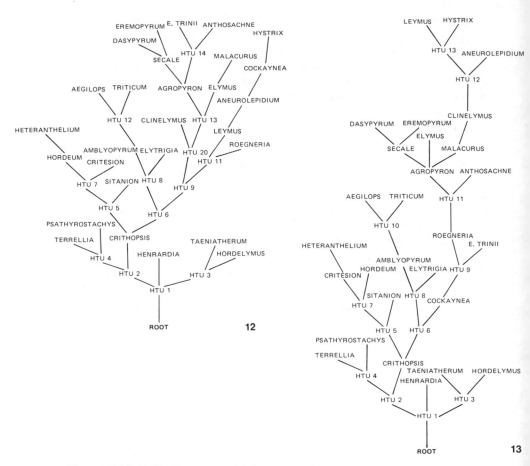

Figures 11.12–11.14. Cladograms of Triticeae genera: trees obtained from series of successive compatibility analyses based on clique 1.

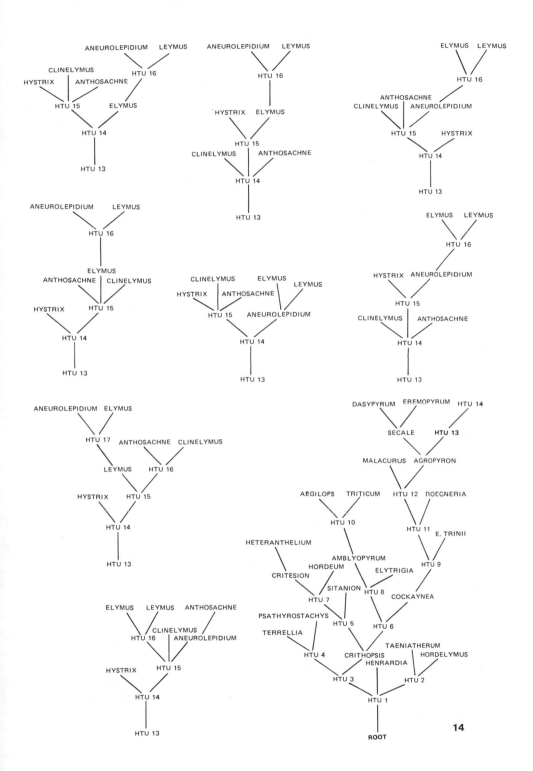

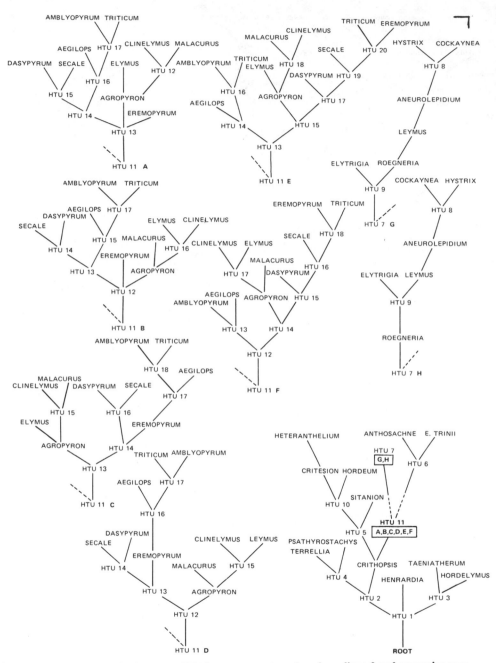

Figure 11.15. Cladograms of Triticeae genera: trees based on clique 2 and successive compatibility analyses.

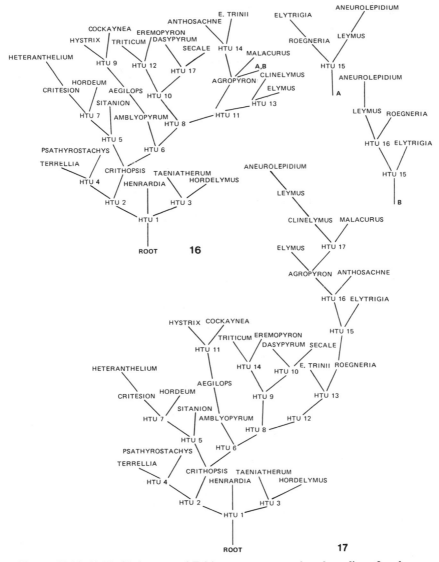

Figures 11.16, 11.17. Cladograms of Triticeae genera: trees based on clique 3 and successive compatibility analyses.

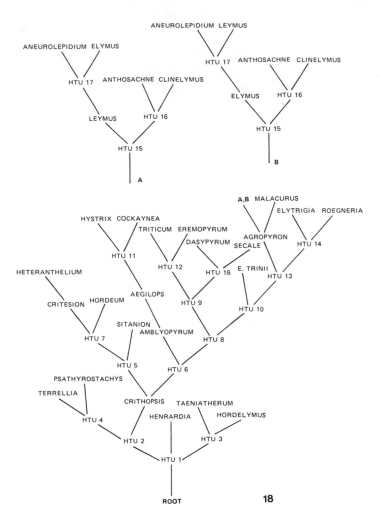

18

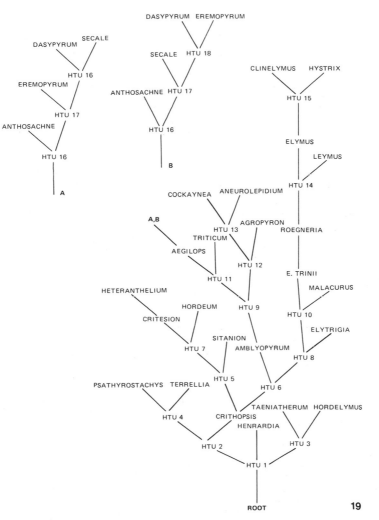

Figures 11.18–11.19. Cladograms of Triticeae genera: trees based on clique 4 and successive compatibility analyses.

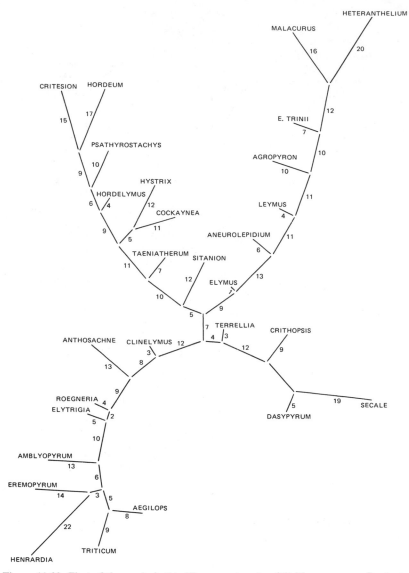

Figure 11.20. First of three admissible Wagner networks of Triticeae genera. See text.

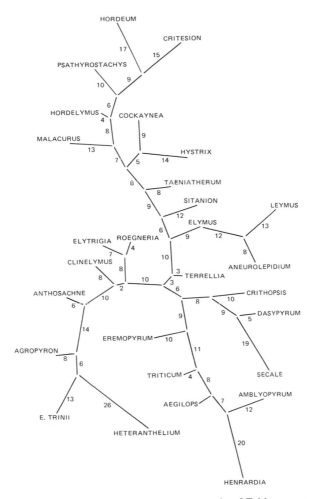

Figure 11.21. Second of three admissible Wagner networks of Triticeae genera. See text.

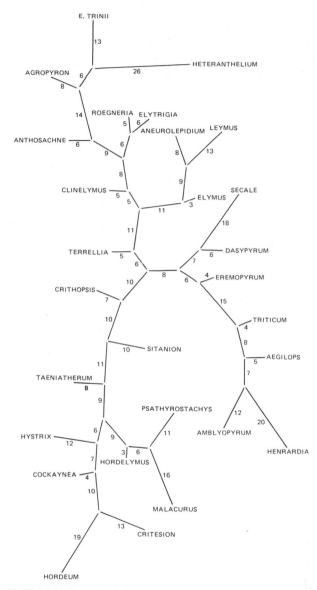

Figure 11.22. Third of three admissible Wagner networks of Triticeae genera. See text.

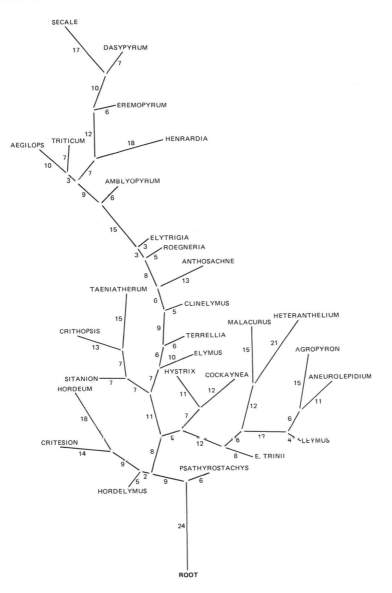

Figure 11.23. Wagner tree with ROOT defined as the vector of all the plesiomorphic states that were hypothesized for character compatibility analyses.

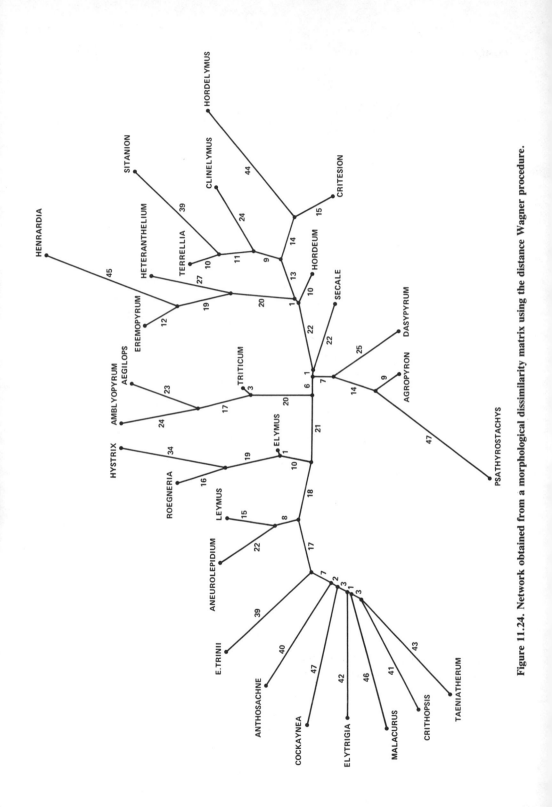

Figure 11.24. Network obtained from a morphological dissimilarity matrix using the distance Wagner procedure.

ful in the interpretation of the results. All the different admissible clado-grams leave us with a number of choices among alternatives. But surely there is room for new approaches to finding optimality criteria.

On the other hand, I am not sure that we need to make a choice at all. This is because the results obtained depend on a combination of factors. It is important to distinguish between the mathematical models, the criteria of fit between the models and the data, and the kind of data coding, e.g., definition of character-state trees. The last is especially important in mor-phological taxonomy. Its importance lies in the fact that morphological tax-onomy is based on sensory data of some kind. The various cladistic tech-niques are used to display those sensory data in a geometrical form of cladistic relationships, thus allowing visual assessments to be made. These assessments, in turn, are important for biological interpretation. In view of this, perhaps we need more than one solution. We may require a number of different outlooks to give us different perspectives on our sensory data input. In spite of their differences, these perspectives are equally informa-tive aspects of the underlying structure (Shepard 1980) we are after—the structure being the true cladogram, which will eventually be unraveled by a series of successive approximations.

Usability

All the programs were relatively easy to compile and link-edit. Slight modifications were necessary for compilation with a FORTRAN-Extended-H compiler and for link-editing on an IBM 3033 running MVS. All the pro-grams use very little CPU time (table 11.?). Of these, CLADON 3 is the least efficient. The two most efficient are WAGNER 78 and CLINCH. The former is the more versatile, in the sense that quantitative data and distance matrices can be processed with it as well.

Conclusions

At least one-third of the species of vascular plants are of polyploid ori-gin. Unfortunately, none of the numerical cladistic methods in current usage accommodates reticulate evolution. But this is not as serious as it seems. At the species level, once information on groups of species with similar chromosome number or with similar genomes is available, numerical clad-istic analysis can be most profitable. Furthermore, it may be the only means of inferring cladistic relationships within these groups, due to limitations of

Table 11.2. Examples of CPU time used by various cladistic programs

Study Collection	Number of EUs	Program	CPU time (minutes)
Avena cultivars	16	COMPATIBILITY	0.061
Diploid Avena species	14	CLADON 1 CLADON 2 CLADON 3	0.004 0.030 0.175
Diploid Avena species	14	CLAD/OS	0.014
Hexaploid Avena species	8	CLAD/OS	0.010
Triticeae genera (primary undirected trees)	28	CLINCH 3	0.006
Triticeae genera (primary directed trees)	29	CLINCH 1	0.005
Triticeae genera (character X OTU matrix)	29	CLAD/OS	0.048
Triticeae genera (distance matrix)	28	WAGNER 78	0.003

Note: See text for details.

the conventional cytological methods. Below the species level reticulate evolution is probably much more common, so that the use of the current numerical cladistic methods is prohibitive. Consequently, we are restricted to particular cases that are suitable to the model. Above the species level reticulate evolution is rare. The current methods of numerical cladistic analysis are more appropriate for EU collections of this kind.

The user of compatibility methods must be aware that only a subset of characters is used to make cladistic inferences. It might be profitable to carry out cladistic studies that would combine morphological data with molecular data in order to help us comprehend the relationship between genetic change through time and concurrent alteration of the phenotype.

Lack of criteria makes it difficult to compare results obtained by different methods applied to the same data. It is, however, always possible to compare the results by visual inspection and by neural assessment—valid actions from the user's point of view. The onus remains with the taxonomist to make a thoroughly responsible choice of objects, characters, character states, and character-state codes, and of the method itself. This responsibility and the taxonomist's interpretation of the results determine to a great extent the relevance of any study.

Literature Cited

Bartcher, R. L. 1966. FORTRAN IV program for estimation of cladistic relationships using the IBM 7040. Computer contribution 6. Lawrence: State Geological Survey, University of Kansas.

Baum, B. R. 1975. Cladistic analysis of diploid and hexaploid oats (*Avena*, Poaceae) using numerical techniques. *Can. J. Bot.* 53:2115–2127.

Baum, B. R. 1977. Oats: Wild and cultivated. A monograph of the genus *Avena* L. (Poaceae). Ottawa: Supply and Services Canada, Ottawa.

———. 1978. Taxonomy of the Tribe Triticeae (Poaceae) using various numerical techniques. II. Classification. *Can. J. Bot.* 56:27–56.

Baum, B. R., and G. F. Estabrook. 1978. Application of compatibility analysis in numerical cladistics at the infraspecific level. *Can J. Bot.* 56:1130–1135.

Baum, B. R. and L. P. Lefkovitch. 1973. A numerical taxonomic study of phylogenetic and phenetic relationships in some cultivated oats using known pedigrees. *Syst. Zool.* 22:118–131.

Baum, B. R. and B. K. Thompson. 1970. Registers with pedigree charts for cultivars: Their importance, their contents, and their preparation by computer. *Taxon* 19:762–768.

Camin, J. H., and R. R. Sokal. 1965. A method for deducing branching sequences in phylogeny. *Evolution* 19:311–326.

Estabrook, G. F. and K. L. Fiala. 1976. COMPATIBILITY. A program to perform cladistic character compatibility analysis. University of Michigan, Ann Arbor. Mimeo.

Estabrook, G. F., C. S. Johnson, and F. R. McMorris. 1976. A mathematical foundation for the analysis of cladistic character compatibility. *Math. Biosci.* 29: 181–187.

——. 1976b. An algebraic analysis of cladistic characters. *Discrete Math.* 16:141–147.

Estabrook, G. F. and C. A. Meacham. 1979. How to determine the compatibility of undirected character state trees. *Math. Biosci.* 46:251–256.

Estabrook, G. F., J. G. Strauch, Jr., and K. L. Fiala. 1977. An application of compatibility analysis to the Blackiths' data on orthopteroid insects. *Syst. Zool.* 26:69–276.

Farris, J. S. 1970. Methods for computing Wagner trees. *Syst. Zool.* 19:83–92.

——. 1972a. Estimating phylogenetic trees from distance matrices. *Amer. Natur.* 106:645–668.

——. 1972b. CLAD/OS. SUNY at Stony Brook, New York. Computer tape.

——. 1978. WAGNER 78. SUNY at Stony Brook, New York. Mimeo.

Fiala, K. 1980. CLINCH 3. University of Michigan, Ann Arbor. Mimeo.

Fiala, K. and G. F. Estabrook. 1977. CLINCH 1. University of Michigan, Ann Arbor. Mimeo.

Gringas, T. R. and R. J. Roberts. 1980. Steps toward computer analysis of nucleotide sequences. *Science* 209:1322–1328.

Grant, V. 1971. *Plant Speciation.* New York: Columbia University Press.

Rajhathy, T. and H. Thomas. 1974. Cytogenetics of oats (*Avena* L.). Miscellaneous Publications of The Genetic Society of Canada, no. 2. Ottawa, Ontario.

Rohlf, F. J. 1974. Methods of comparing classifications. *Ann. Rev. Ecol. Syst.* 5:101–113.

Shepard, R. N. 1980. Multidimensional scaling, tree-fitting, and clustering. *Science* 210:390–398.

Sneath, P. H. A. 1975. Cladistic representation of reticulate evolution. *Syst. Zool.* 24:360–368.

Stanton, T. R. 1955. Oat identification and classification. Technical Bulletin no. 1100, U.S. Department of Agriculture, Washington, D.C.

Stebbins, G. L. 1966. *Processes of Organic Evolution.* Englewood Cliffs, New Jersey: Prentice-Hall.

12

Cladistic and Other Methods: Problems, Pitfalls, and Potentials

WALTER M. FITCH

Introduction

I am supposed to talk about matrix methods, and I shall, as well as give an overview of the material you have heard. Matrix methods are phenetic methods. My best-known paper (Fitch and Margoliash 1967) presents a phenetic method that is closely related to the unweighted pair-group method of analysis (UPGMA), although it has several features which distinguish it from UPGMA. Hence, there are many who think of me as a pheneticist. Apparently, after some of my questions earlier in this workshop, there were some here who had tagged me as anticladist. Nothing could be further from the truth. I have spent the last decade promoting parsimony techniques but I have been asked to discuss matrix methods. I even have a new matrix method which is, I think, quite different from anything you have ever seen before. It is called "neighborliness" and is based upon the additive properties of the four-point metric (Fitch 1982). It had been my intent to disucss this method. But I am not going to do that because, having listened to the talks and discussions at the workshop, I thought maybe it would be more useful for you as students here not to be plagued by yet another method but rather to have me, as the last speaker, make some additional remarks that may seem a little bit disjointed at times but which fill in what I perceive as some gaps in this week's education. I hope this will make you a bit more cautious about the truth of any one method's results and yet more adventurous in trying new methods.

First of all, am I a cladist? I think I would prefer not to have any title. I

This work was supported by a grant from the National Science Foundation, DEB 78-14197.

don't see in my mind that what I say in any given instance is true or false depending on whether I am a cladist or not. What I say can best be judged by you for its correctness independently of whether or not I am a cladist. Indeed, it is my view, which I hope to develop here, that there are good reasons for us to be eclectic in our philosophies.

Why Parsimony?

Before I get to the matrix methods, I would first like to ask the question, Why parsimony? Consider a single character, an amino acid, which is different in each of four taxa. Figure 12.1 shows, in the upper part, that character, and the upper right-hand matrix presents the distance between each pair of taxa as one. This is called additive data because you can fit a tree with lengths upon the branches, as I have done in figure 12.1, for which the sum of the distances between any two taxa along their connecting branches is exactly equal to the distance in your observed difference matrix. So you can see we have created the perfect solution. What is interesting, however, is that I could have chosen any of the three possible unrooted relationships of four taxa and the optimal solution would still have been additive for them. Thus additivity is not in itself the end of everything you would like. Additivity is a very comforting property for the data to have but is something almost never found in real data.

Now, pairwise amino acid differences are what, for example, immunological techniques attempt to approximate, except that they are summing over more characters than just one. Every time there is an amino acid difference between two sequences, you would like the immunological technique to detect that difference. To the extent that it fails to, the immunological technique is only an approximation of this kind of data. As the discussion develops in the four cases in figures 12.1 and 12.2, you will see why one might worry about difficulties with immunological data. Notice that there is a length to the tree. The length is two; that is the minimal length of the tree. Since the data are additive, I should be able to find an additive solution, and that is what is shown.

Now, one would like to analyze the problem more deeply. Notice that since I always deal in molecular sequences, I do not have an ordered transition series for those states. That is, I cannot say anything about which state is primitive and which is advanced, but there are steps that we can take that correspond crudely to additive binary coding. We can use the genetic code to say that these differing amino acids are not all different to the same extent. Any pair of different amino acids is only one amino acid difference,

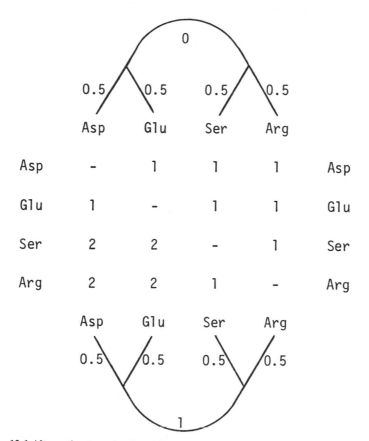

Asp	–	1	1	1	Asp
Glu	1	–	1	1	Glu
Ser	2	2	–	1	Ser
Arg	2	2	1	–	Arg

Figure 12.1 Alternative trees for four different amino acids, one found in each of four taxa, based on absolute difference and the genetic code.

but the genetic code says that some of these amino acids have two nucleotide differences. Thus, using the genetic code, we get a different matrix of distances from what we had before (shown in the lower left half of figure 12.1). And you can see we have an increase in information. Before, all taxa were equally distant. That is no longer true. I chose the particular topology in the upper half of figure 12.1 so as to match the optimal topology for the data in the lower half to show the progress that we make in these matrix techniques as information is added. We have the same tree topology, we have numbers on its legs, and again the data are additive. They are, however, additive only for this topology, not for either of the other two. Notice the tree is now one unit longer than before (length = 3), so it is clear we have gained information by using the genetic code.

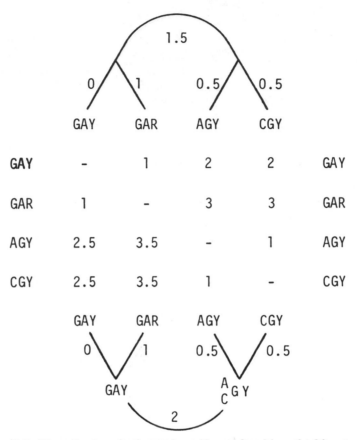

Figure 12.2. Alternative trees for four amino acids, one found in each of four taxa, based on minimum base differences and codon differences.

In figure 12.2 we see the same data again, but in the upper right half we have made a further improvement. The minimum base differences were two between the two pairs of taxa in the previous example, but now you see two twos and two threes as their difference. How did we get that increase? The minimum base difference between arginine and aspartate can be two, provided the arginine codon has a pyrimidine in the third position. It can be two between arginine and glutamate also, provided, however, that the arginine codon has a purine in the third position. But you cannot have it both ways; either it is a purine or it is a pyrimidine. (Heterozygosity does not really help, for then we just assume the extra change occurred within the arginine codon in order to produce the heterozygosity.) In the upper part of figure 12.2, I've represented the amino acids by their codon form, where Y and R are pyrimidines and purines, respectively, and the choice of an arginine codon forced the result shown. (An equivalent result would have occurred if the arginine codon selected had been AGR rather than CGY.)

The result is that we have a third set of data that obviously has still more information in it. This increase is the result of restricting the transition states, much as you do when you do additive binary coding. We do not always achieve the full information that you are expected to do with your techniques, but the analogy is correct. You will see that this tree result is still additive, that it is a perfect reconstruction of the observational data, if you will add the numbers on the tree. You can also see that the length has now increased to 3.5. It was 2, then it was 3, now it's 3.5. All of these are lower bounds, the minimum length that the tree must have to explain the data in the matrix. We are increasing these lower bounds.

All of the above are matrix procedures. We now go to the solution in the lower half of figure 12.2, in which the matrix is irrelevant because the matrix was filled in after I got the tree rather than being used to construct it. I did a most parsimonious reconstruction, always choosing codons to minimize the total amount of change on the tree. They are the same as those in the upper tree. Of course, that was deliberate. You can see that this is the longest tree yet (length = 4). This is a most parsimonious tree. There is no criterion of goodness of fit or additivity, only that the change be as little as possible. But it is longer than the others. It is a proposition in simple logic that if you have a series of estimates all of which are lower bounds to the true historical set of changes, the lower bound that is largest is closest to the truth. Thus, the parsimony procedure has found more of the historical change than any of the three matrix procedures.

That this tree is most parsimonious does not mean that historically it necessarily had to happen this way. It says an aspartate codon changed to a glutamate codon by virtue of some pyrimidine (we do not know whether it is C or U) changing to some purine (we do not know whether it is G or A). But there had to be at least one change. There had to be such a change to account for the data, but it does not say that it could not have gone from C to U to A or from C to A to G. There could be other changes hidden in the data, but as there is no way of knowing what they are, their specification only increases the number of hypotheses required to explain the data. This is not to say that one or more unknown intermediates may not have been highly probable, but at least one event is certain, additional events are uncertain. The largest lower bound is giving you the most information about the tree structure. The largest lower bound is a parsimony reconstruction. It is always at least as large as any of the other kinds of matrix methods built upon data like these or it should be if all of the methods that are being used are in some sense minimal-estimate techniques. That is because you lose a lot of information contained in the character-by-taxon matrix when you summarize it by some kind of averaging process into the set of numbers in the distance matrix of a phenetic method.

Why Matrix Methods?

If parsimony is so much better, why do we have matrix techniques? There are a number of reasons, the most outstanding of which is probably that some data only come that way. It is the only way you get immunological data, and you cannot go back from there to something more definitive in terms of character state. DNA hybridization, looking at melting-point lowering, is another kind of information that comes to you as pairwise distances. There is a need for techniques for such data, and it is with such a need in mind that Steve Farris (1972) developed the Wagner distance program. Also, it can be useful to give you some rough idea of what is going on in your data before you do the long and hazardous labor that is necessary for a good cladistic analysis. Indeed, you earlier heard Wagner (this volume) advocate that you should do some kind of phenetic analysis first.

Why Do Matrix Methods Fail?

What is the nature of the source of our problems that all methods do not give you the same right answer? In figure 12.3 you see two data sets divided by the series of asterisks. Focus first on the left, where we see a hypothetical evolution of an ancestral sequence of 22 As into four descendent sequences after 22 mutational steps that change As to Cs at the point where a C is overlined. It looks very much like what Dan Brooks (this volume) showed you earlier in the workshop. The subscript denotes a set of characters all of which have equivalent state changes. Now what was done here, with great deliberateness, was to allow no character to change its state more than once. You may compare any two sequences to see how many differences there are (say between gamma and delta, which differ in eight characters), as shown in the matrix. If you examine the tree, you will see the number of changes between them on the tree is the same (eight). Because we never had any character change more than once, these data are additive; that's why additivity is interesting. This means that if I now ask what is the best tree I can reconstruct from the difference data on a topology like the one shown, I will correctly place the appropriate numbers on the proper legs. In other words, if you compute the phyletic distances, those distances that you get by summing the lengths of the legs of the reconstructed tree, they add up to the number of differences. This only occurs when each character only changes once. (Alternatively, if you have a character change more than once, but each time to a new state where the order of the character-state changes is known or imposed, you can still have additivity with multiple changes in the state.)

Homoplasy

What happens when characters change more than once? Examine the right-hand side of figure 12.3. We have reduced the central part of the sequence from eight to four positions and placed all eight of their left-handed changes into these four, so that there are four parallel A → C changes in the descent to α and δ. We have put homoplasy into the data. When you look at pairwise differences you see that the numbers are very similar. The only difference is between the two taxa that suffer the four parallel changes. There are now 8 differences that have disappeared, so that now, instead of the differences between α and δ being 22, they are only 14. This makes the data nonadditive and causes problems when you use matrix techniques. Most matrix techniques, Wagner distance (Farris 1972) being an exception, have some kind of implicit uniform rate feature to them. The conclusion is that beta and gamma, which differ by only 6, are more closely

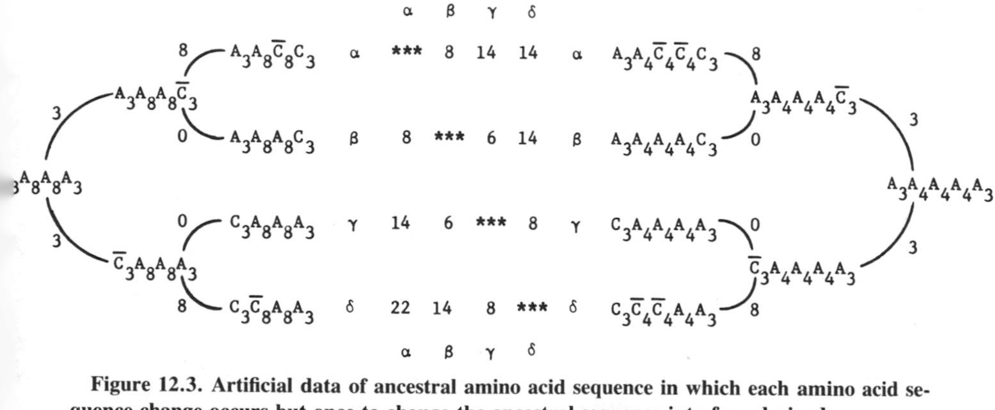

Figure 12.3. Artificial data of ancestral amino acid sequence in which each amino acid sequence change occurs but once to change the ancestral sequence into four derived sequences.

related, with the result that they are joined together as sister taxa because their differences are less than those between other pairs of taxa. This is most likely to occur, and be wrong, as Felsenstein (1973) showed, when you have substantial inequalities of evolutionary rates. That is the case in our example, where, from their common ancestor, one line of descent changes eight times, the other not at all. That is the nature of the problem introduced by matrix methods. That is why cladistic analysis, when you can perform it, is more likely to give you a satisfactory result.

To see how UPGMA gets the wrong tree, examine figure 12.4, which shows, on the far left, the so-called truth. But we cannot know about that, as we only see the matrix of pairwise differences shown in the middle. If you do a UPGMA on the middle data matrix, you get the result shown in the upper right tree, with the corresponding phyletic distances shown in the upper right half of the matrix below it. Notice that it has the wrong topology. The Fitch-Margoliash method (1967) gets the result shown in the lower right tree, with its phyletic distances shown in the lower left of the matrix immediately above it. It has a considerably lower percent standard deviation (%SD). The Wagner distance method gets the lower right tree also, except that the internal branch would have a length of six.

EXAMPLE OF EFFECT OF PARALLEL CHANGES

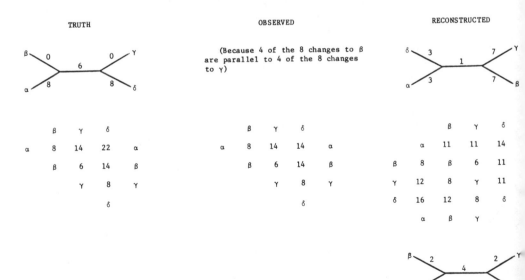

Figure 12.4. Artificial data used in figure 12.3, with parallel changes added to illustrate their effect on tree topology.

Unequal Rates

Now let me say something about the assumption in phenetic methods of uniformity rates. It is not absolute, but if the joining of two taxa is done because they are closer than any other as yet unjoined taxa, then a relative uniformity is implicit even if sister taxa are not assigned legs of equal length. For example, figure 12.5 shows the data from which you are to create a tree. If you say, "Hey look, the bird and the turtle only differ by 11," and join them, you create the lower tree, where the phyletic distances precisely equal the observed values of 11, 55, and 58. There is your tree, and be-

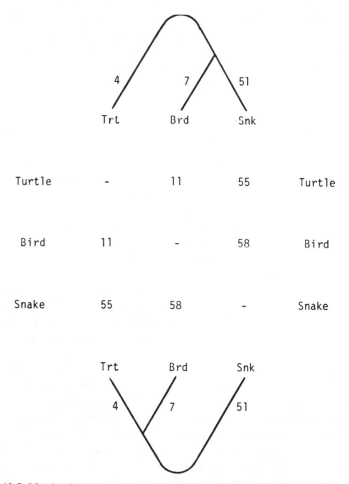

	Trt	Brd	Snk	
Turtle	-	11	55	Turtle
Bird	11	-	58	Bird
Snake	55	58	-	Snake

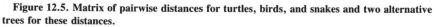

Figure 12.5. Matrix of pairwise distances for turtles, birds, and snakes and two alternative trees for these distances.

cause the outgroup is so far away you put a root down there at the bottom.
The fact that I drew a semicircle is to indicate that I am not overly keen on
placing the root at that point. The root is not an important feature as far as
parsimony is concerned.

On the other hand, what happens if we make an alternative choice? We
can decide to be biological and put the birds with the snakes since they
are considered more closely related by biologists. Now it turns out that we
have only three taxa and thus we can always fit the data perfectly. And so
you see the legs in the upper tree have the same lengths. However, there
is no unweighted pair-group method that is going to get you that result,
despite the fact that the numbers are the same. It is going to force the lower
tree upon you. Thus, while the Fitch-Margoliash procedure (1967) does not
assume equality of rates, the choice as to which joinings to make and the
order in which they occur are a function of relative differences. This amounts
to an implicit assumption of relative equality of rates, and that is another
reason why matrix methods have some deficiencies.

Minimal-Length Trees

In figure 12.6 we see, in the lower left corner of the matrix, Sarich's immunological data for the carnivores (1969). The monkey was used as the
out-group. The data were analyzed by my neighborliness method (Fitch
1982), and it gets the same topology as Farris (1972) did when he used the
data to exemplify his Wagner distance program. The only difference is slightly
different numbers on some of the branches so that the total length of my
tree is three units shorter than Farris' tree and is the true minimum length
he was trying to find. However, that is an irrelevancy. Rather, I bring this
up to show some properties of minimal-length trees that we have proven
(Fitch and Smith 1982), properties that are implicit in any program like
Wagner distance that requires the sums of the distances on the legs connecting any two taxa to be at least as large as the observed pairwise distance. The phyletic distance can never be less than the observed difference. The phyletic distances are shown in the upper right half of the matrix.
Given that condition, if you then minimize the length of the tree, several
conclusions necessarily follow. First of all, for any two taxa that are separated by a single node (sister taxa), the observed distance must equal the
phyletic distance. Moreover, any interior node divides the tree into three
subtrees and there must be at least one pair of taxa for each of the three
pairs of subtrees with equal observed and phyletic distances. For example,
for the node that divides the tree into the cat, the monkey, and all others,

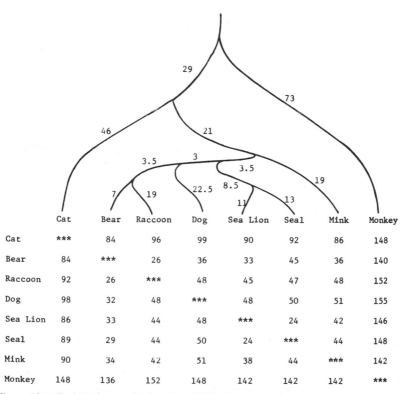

A PHYLOGENY OF CARNIVORES BASED ON SARICH'S
IMMUNOLOGICAL DISTANCES

	Cat	Bear	Raccoon	Dog	Sea Lion	Seal	Mink	Monkey
Cat	***	84	96	99	90	92	86	148
Bear	84	***	26	36	33	45	36	140
Raccoon	92	26	***	48	45	47	48	152
Dog	98	32	48	***	48	50	51	155
Sea Lion	86	33	44	48	***	24	42	146
Seal	89	29	44	50	24	***	44	148
Mink	90	34	42	51	38	44	***	142
Monkey	148	136	152	148	142	142	142	***

Figure 12.6. Sarich's immunological data (1969) for carnivores, with the trees obtained by the neighborliness method (Fitch 1982) and the Wagner distance method (Farris 1972).

the observed and phyletic distances from monkey to cat are 148, those from monkey to mink are 142, while those from cat to bear are 84.

Finally, the most powerful result of all. As you let your eyes travel down the semimajor diagonal immediately above and below the asterisked diagonal, you will see the identical series of numbers, 84, 26, 48, . . . etc. This means that for any two taxa next to each other at the tips of the tree, the phyletic and observed distances are equal. It is also (and should be) true for the two taxa at the ends (monkey and cat). Granting the ability to rotate parts of the tree around any interior node without changing the tree, we assert that if it is possible to bring about this arrangement of equalities, then the tree is truly minimal. We have proven it for as many as five taxa and it has always been true for larger numbers of taxa, as in the carnivore data of figure 12.6. McMorris and Neuman (1983) have proven this for the general case by showing that the condition is sufficient for minimality, while Hendy (1983) has shown that the condition is not necessary if there are more than two pairs of sister taxa in the tree.

An implication of the condition that the phyletic distance equals the observed distance is that there is no detectable homoplasy between any of those neighboring pairs. There is no added distance beyond what we observed in the first place. This property, a property of all minimal-length trees, is peculiar because it says we will not permit the homoplasious steps required by the nonadditivity of the data to occur in certain relationships. In particular they are not to occur between the out-group (with which the most homoplasy ought to be associated) and at least two of the taxa that are in the group that you are examining. In figure 12.6, this means no homoplasy between the monkey and the cat and mink. This is a somewhat unhappy restriction on the distribution of those homoplasious events demanded by the data.

I would like to make it clear that this condition is no objection to using the method to find the topology. Finding the topology which has the minimum length is still pertinent, it seems to me. What is more tenuous is to take the numbers on the branches and infer anything about evolutionary rates. In addition to the lack of homoplasy between sister taxa, I have yet to find any real data for which a minimal tree could not give you an infinity of alternative branch values whose sum was still the same minimal length, even while maintaining the constraint that all the neighbor taxa have observed distances equal to phyletic distances. Thus, I would say that one ought not to be too hasty to make inferences about rates of evolution from such trees.

Maximum Parsimony—Some Problems

Optimizing the Character States on the Tree

That concludes what I wish to say about matrix methods in general. I would now like to make a few general remarks on parsimony. The first is just a plug. You heard earlier about optimizing ancestral states on a tree by the "up-down routine." Recall that the problem is, given the character states of the taxa and the topology of the tree, what is the minimum number of state changes. My method (Fitch 1971) is more general because:

1) it does not require you to know where the root is;
2) it does not require you to know anything about what states are primitive or advanced;
3) it does not require you to limit yourself to two-state characters (however, if there are multistate characters, the assumption in the algorithm is that each character state can go to any other character in one unit of change)
4) it does not require every taxon's state to be completely defined (this last is accomplished by defining the character state of a taxon as the set of all states you do not wish to exclude); and
5) all possible equally parsimonious solutions are obtained.

If you want a rooted tree, simply add a "hypothetical" ancestral taxon in the appropriate location with the appropriate character states assigned, primitive where known, ambiguous where unknown. If you do not wish all states to be one distant from every other, additive binary coding immediately resolves that difficulty so that all cladistic analyses are special cases of my general solution.

Adding One Character Can Change the Tree

I would like to make several points. The first point is that the addition of one new character can change the topology of the most parsimonious tree. In table 12.1 you see the states for six characters and five taxa. If one ex-

Table 12.1. Hypothetical data matrix for five taxa and six characters

Taxon	Characters and States					
	1	2	3	4	5	6
alpha	A	A	C	C	A	A
beta	A	A	C	A	A	C
gamma	A	C	A	C	A	A
delta	A	C	A	A	C	C
epsilon	C	A	A	A	A	A
zeta	C	A	A	A	C	C

amines only the first five characters, the tree structure will be determined by the first three characters, the third and fourth being designed to introduce just enough discord that the addition of the sixth character tips the balance to a radically different tree. Figure 12.7 shows the most parsimonious tree for five characters at the bottom (seven changes of state) and for six characters at the top (nine changes of state). There is only one parsimonious tree in each case, and it is the same regardless of whether no state is designated as ancestral (primitive) or A is designated as ancestral.

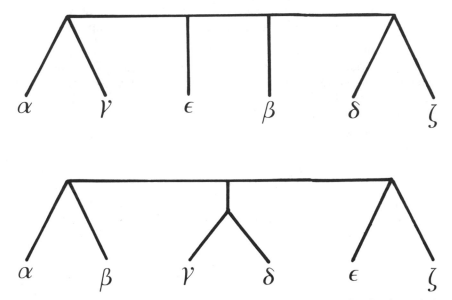

Figure 12.7. Alternative topologies for the data in table 12.1, illustrating the change in topology caused by the addition of one character.

Adding One Taxon Can Change the Tree

The second point is that the addition of a single taxon (say zeta) can alter the structure of the most parsimonious tree. That is, if one simply removes that taxon from the tree, the tree that remains is not the topology of the most parsimonious tree for the remaining taxa. Table 12.2 shows the data for five characters and six taxa. If only the first five taxa are examined, the most parsimonious tree (eight changes of state) has the topology shown in the upper portion of figure 12.8. If all six taxa are examined, there are two most parsimonious trees (nine changes of state). One of them is shown in

Table 12.2. Hypothetical data matrix for six taxa and five characters

| Taxon | \ Characters and States | | | | |
	1	2	3	4	5
alpha	A	A	C	A	A
beta	A	C	C	C	C
gamma	A	A	A	A	A
delta	C	C	A	A	A
epsilon	C	A	A	C	A
zeta	C	C	A	C	C

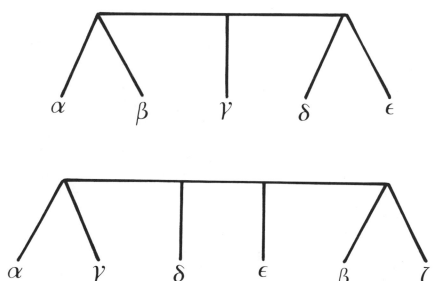

Figure 12.8. Alternative topologies for the data in table 12.2, illustrating the change in topology caused by the addition of one taxon.

the lower half of figure 12.8; the other is identical except for the inter-change of taxa delta and epsilon. Removal of zeta would not produce the upper tree in either case.

The Maximal Clique Tree May Be Inconsistent with the Most Parsimonious Tree

The third point is that the most parsimonious tree is not necessarily con-sistent with the tree for the maximal clique. In table 12.3 are shown seven characters for six taxa, but imagine that each of the first six characters is really representative of three different characters that all give equivalent in-formation (hence their multiplicity equals three), while the seventh is four equivalent characters. Since all seven of the characters shown are pairwise unavoidably discordant or incompatible, the maximal clique is composed of the four equivalent representations of character 7. The most parsimon-ious tree for the maximal clique requires that some branch of the tree par-tition the taxa into two groups, one containing α, γ, ϵ, the other β, δ ζ. There are six most parsimonious trees for the complete data set, and none of them is consistent with the data for the maximal clique.

Table 12.3. Hypothetical data set of six taxa and seven characters, each of which represents a set of characters with identical character-state trees and distribution among the six taxa

Taxon	Characters and States						
	1	2	3	4	5	6	7
alpha	A	A	A	A	A	C	A
beta	A	A	A	A	C	A	C
gamma	A	C	C	C	A	A	A
delta	C	A	C	C	A	A	C
epsilon	C	C	A	C	C	C	A
zeta	C	C	C	A	C	C	C
multiplicity	3	3	3	3	3	3	4

Outgroups for Primitiveness and Rooting Trees

Let us return to point two, the adding of a new taxon, for which we have two principal reasons. One reason is we want to have an out-group. This is common and reasonable to me. A second reason is that we want a hypothetical taxon which simultaneously serves two purposes, the rooting of the tree and the defining of the primitive state. In fact you have been doing that routinely in your cladistic analysis. When you add a taxon with all states equal to zero, you are adding a hypothetical taxon and asserting that zero is the primitive state. Now that hypothetical taxon can affect what the most parsimonious tree is. And perhaps it should. Nevertheless, it is possible, as we have seen, that the most parsimonious tree for the taxa without the hypothetical taxon would show a different relationship (the addition of a taxon can change the tree). That raises then the question in your mind, or should, "Do I really want that?" The answer is, like most things, it depends. We have a situation such as David Wake (1981) described in this workshop as "either reciprocal illumination or, sometimes, reciprocal self-deception." The answer lies in determining which of two related pieces of information is most secure. Initially we have character states for the characters of taxa. It is our task to discover the evolutionary changes in the characters (Stuessy and Crisci, this volume) and the phylogenetic relationships among the taxa. These are the two interrelated pieces of information sought. It should be clear that if we knew one of these for certain, the task of inferring the other would be considerably eased. If neither is known for sure and you wish to estimate both, then you should accept the piece of information in which you have the greatest confidence and use it to estimate the other piece. This was one of the important concepts David Hull (this volume) presented to you earlier. If you know for certain all the primitive states, it would be a mistake not to put that hypothetical taxon with all its primitive states into your analysis. It would be a mistake, because that information is giving you all kinds of help. But what if you do not know a thing about primitiveness? You know nothing. It would be an equally gross mistake to put that hypothetical taxon in. You should do the analysis without that hypothetical taxon.

What is the realistic case? Neither of those extremes, usually. It is more often the case that for some of your primitive states you have a great deal of confidence and for others you have very little confidence at all. Is it possible to incorporate this in your data analysis? Certainly; that is what my generalized parsimony procedure (Fitch 1971) does for you. Remember the fourth general characteristic of that process, which was that any degree of character-state ambiguity is allowed. Thus, for any character of whose primitive state you are confident, that state is assigned to the ancestral taxon, whereas for any character whose primitive state is uncertain, an ambiguous state is assigned to the ancestral taxon. If it is a multistate character, one can even assign to the ancestral taxon a subset of the possible states that excludes all states you know cannot be primitive.

Weighting Characters

Deciding whether you know a character well enough to decide its primitive state is akin to another problem, the weighting of characters. Weighting characters has tended to be anathema to people because it seems so difficult to figure out how to weight and you feel it is better to weight all the characters equally. I have a great deal of sympathy for that because I have been thinking for a long time of how to weight them and I am not sure that I have any answers. I will make one suggestion at the very end here, but I think that first you ought to disabuse yourself of the thought that you may not be making a judgment about weights. Nothing could be further from the truth. In the so-called absence of weighting, you are saying the weight of this character is 1 and the weight of this character is 1 and this character . . . etc. (I can do this ad nauseum if you need to be impressed at the fact that you're weighting.) And if you had 83 characters and you decided not to use 37 of them, then you have already invoked unequal weighting, because for those other 37 you said "weight equals 0, weight equals 0, weight equals 0, etc." You have already induced a weighting system. The only question is, how far are you going to go with it? Well, however far you wish, but let me point out an advantage to putting weight into the system at the outset.

What do you do with equivalent characters? If two characters are giving you the same information because they have the same character-state tree, it sometimes occurs that the taxonomist keeps only one of them because you only need one good synapomorphy to support the tree, right? But if you just keep one of each equivalent character and you keep looking for more characters, you're going to find characters which refute every evolutionary hypothesis. And if that is so, then you obviously had better have at least two synapomorphic characters so that one other character doesn't refute it, because all the synapomorphies, as they accumulate, make it that much harder for random noise among those which are not good characters to overturn the appropriate tree. Does that require more computer time? Not really. You can just keep your one character of each type if you set the weight equal to the number of characters that have that equivalence relationship. If you have five characters that all have the same character-state tree for the taxa, then just give a weight equal to five for one exemplar of that character instead of a weight equal to one, and whatever is the result of the analysis, just multiply it by five and add it in. That saves time and makes the program more efficient.

General Remarks

An Answer Ordained

I now want to make some remarks on methods in general. The first is, best fit methods produce an answer however bad the data. Now it is unfortunate, but output seems to lead people to believe that they have truth. You are going to get a tree whether you like it or not. I gather a lot of times during this workshop, you have not liked it.

Criteria Should Be Appropriate

Every tree-building method should have an appropriate criterion, and they are not all the same. My method (Fitch and Margoliash 1967) uses %SD. That is appropriate to that method, but it is not appropriate to a Wagner distance program, which has a criterion that the best tree topology is the one that gives the minimum length. That is appropriate to that program. It is not appropriate to an unweighted pair-group method. You cannot compare the outputs of different methods by the criterion of one method. For example, if you use my method as published in *Science* and use as the standard the length of the tree, which is the Wagner distance criterion, my tree will always win and that's baloney, because the conditions by which I construct my tree are such that they have different underlying constraints than does Farris' program. So the second remark is that you cannot necessarily use one criterion for trees produced by different methods.

Methods Should Examine Alternative Trees

The third remark is that the tree-building method, if it is going to be any good, must examine alternatives. There may be many unrooted trees. If you are a chemist and you have more than 22 taxa, you have more than a mole of tree alternatives. The number of unrooted trees is the product of successive odd numbers up until $2t - 5$. That obviously goes up very fast if you have nearly a mole of them with only 22 taxa and many more if you have more than 22 taxa. You can see you are not going to examine all of those trees with any computer. I know of no methods that are in common use that in fact get you, on the first pass, the best tree by the criterion that is being used. Whatever the criterion being used, you should at least look in the neighborhood of your first tree to see if there is another tree that is better. You ought to ask yourself about any proposed method whether it looks for alternatives and to what extent you are able to ask it to explore the neighborhood and how.

There are various ways to explore the neighborhood. Branch swapping, for example, is one. For any tree you may ask, "If we swap these two neighboring branches, will we get a better tree?" That's one. Another is to ask, for each joining of two sets of taxa you choose to make, whether there is another alternative that doesn't seem to be too much different. If so, then go back and try that one instead. Programs should do that.

So Many Trees and So Little Difference

My fourth remark is that we ought to be asking people who publish these trees to tell us how many trees were examined and how much worse was the second-best tree. Most parsimonious trees are really neat. I am all for them, but have you any idea about the distribution of all the trees in the universe that you might have sampled? In case you haven't, the next few figures will show some distributions.

Figure 12.9 shows the distribution for cytochrome sequences from eight bacteria. There are 10,395 possible phylogenetic trees, and they were all examined. There are two distributions of trees in this figure. One distribution (white heights) is looking just at the minimum number of amino acid replacements required of each tree and plotting how many trees require a given number of replacements. The other distribution (hatched heights) is looking at the minimum number of nucleotide substitutions for the same data, showing again that if you do my crude genetic equivalent of additive binary coding, it is helpful. I plotted both distributions about their respective medians of 133 replacements and 175 substitutions. The range of the amino acid replacement distribution is only 41. The worst tree is only 41 replacements worse than the best one. If you know that you are distributing more than 10,000 trees into only 41 categories it becomes a little more difficult to say that a tree down close to the low end is all that much better. You do considerably better once you have converted the amino acid se-

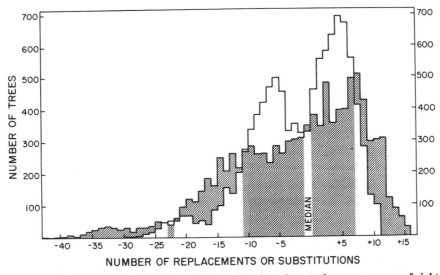

Figure 12.9. Distribution of phylogenetic trees, based on the cytochrome sequences of eight bacteria. Both minimum amino acid replacements and minimum nucleotide substitutions are shown.

quence into a nucleotide sequence. You are, in effect, creating more characters by going to the genetic sequence. You see the range is much greater, from -43 to $+16$ substitutions. The skew of the distribution is very much to the right, which is highly desirable in that this leaves fewer trees in the tail where one hopes to discover the phylogeny.

Figure 12.10 shows the distribution for eight mammalian alpha hemoglobins used by Goodman et al. (1979) to illustrate their ideas for adding gene expression events into the parsimonious counting of events. I used the distribution shown to show how easy it would be to get the tree of your prejudiced choice by invoking gene expression events at a low cost. This is because there were so many trees within the range of the few penalty nucleotide substitutions assessed for each gene expression event invoked. The penalty is added to the more parsimonious trees (making them less parsimonious) because they do not adhere to the tree of your a priori choice, thus permitting you to conclude that your choice is most parsimonious. I have never observed so symmetrical a distribution of tree lengths, and it suggests to me either that the alpha hemoglobins are less than optimal to determine the order of splitting of the major mammalian orders or that they diverged from each other so closely in time that perhaps no sequence will prove adequate for the job. This form is most likely for bushy trees where all lines emanate from a single point.

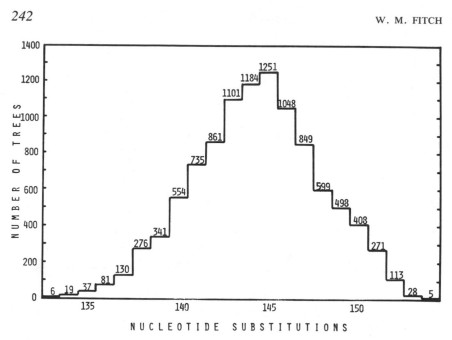

Figure 12.10. Distribution of trees with various numbers of nucleotide substitutions, based on the mammalian hemoglobin data of Goodman et al. (1979) and Goodman, Czelusnick, and Moore (1979).

Figure 12.11 shows the distribution of lens alpha crystalline sequences and represents the most skewed distribution I have so far observed. This form is more likely for the "stringy" trees, those which are maximally asymmetric in their splitting and have long intervals between splits.

Artificial Examples

Next, I would like to present you with some artificial data, devised in a manner I would recommend your trying. Take a single character and define a set of character states for some small number of taxa (at most five), and ask how each method would treat this character on each of several possible tree topologies as if other data might have compellingly argued for each such tree structure. Repeat this for other distributions of the character states. Note how the *difference* in the goodness of fit of the character to different trees is not the same for different methods. This means that by the appropriate choice of the characters to combine and consider all together, one can deliberately create data sets that make different methods yield dif-

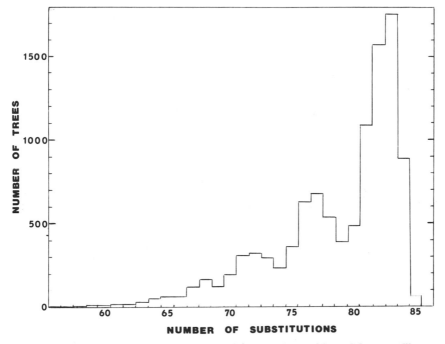

Figure 12.11. Distribution of all trees derived from analyses of lens alpha crystalline sequences, plotting the number of trees that show various numbers of amino acid substitutions.

ferent trees using each method's own criterion. I have done this to get the example shown in figure 12.12.

There are 12 characters of three types. Two, four and six are equivalent characters for Types I, II, and III, and the states are shown for four taxa along with a matrix of pairwise differences. There are only three possible unrooted trees, and the results are shown from the Fitch-Margoliash (F/M) procedure in the next row of data. The labels at the tree tips also serve as column markers for the matrices below. The lower left half of the matrix simply contains the pairwise differences reordered to reflect the tree above. The upper right half gives phyletic distances found by summing the values along the connecting legs of the tree. Notice that the lowest %SD attends the tree that divides the taxa (α, β) and (γ, δ). It also has a negative branch length, to which some object. While the objection has merit, it would be a mistake to believe that, when using any average linkage-type method (UPGMA, F/M), one tree is necessarily less likely to be correct than another simply because the former has a negative branch and the latter does not. There is no "true" answer, since these are made-up data, but I have clearly

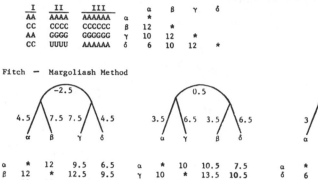

I	II	III		α	β	γ	δ
AA	AAAA	AAAAAA	α	*			
CC	CCCC	CCCCCC	β	12	*		
AA	GGGG	GGGGGG	γ	10	12	*	
CC	UUUU	AAAAAA	δ	6	10	12	*

Fitch — Margoliash Method

		α	β	γ	δ
α		*	12	9.5	6.5
β		12	*	12.5	9.5
γ		10	12	*	12
δ		6	10	12	*

length = 21.5
% SD = 11.7

		α	γ	β	δ
α		*	10	10.5	7.5
γ		10	*	13.5	10.5
β		12	12	*	10
δ		6	12.	10	*

length = 20.5
% SD = 33.1

		α	δ	β	γ
α		*	6	11	11
δ		6	*	11	11
β		12	10	*	12
γ		10	12	12	*

length = 20
% SD = 18.4

Farris Method

Minimum Length Method
(Linear Programming)

	α	β	γ	δ	
α	*	12	12$_2$	6	α
β	12	*	12	12$_2$	β
γ	14$_4$	12	*	12	γ
δ	6	10	12	*	δ
	α	β	γ	δ	

length = 21
% SD = 20

length = 21
% SD = 20

Parsimony Method

	α	β	γ	δ	
α	*	16$_4$	10	10$_4$	α
β	14$_2$	*	22$_{10}$	10	β
γ	12$_2$	20$_8$	*	16$_4$	γ
δ	8$_2$	12$_2$	14$_2$	*	δ
	α	β	γ	δ	

None of II central
length = 26
% SD = 41.6

One each of II central
length = 26
% SD = 58.3

Single Linkage Method

γ —10— α —6— δ —10— β

length = 26
% SD = 62.9

	α	β	γ	δ
α	*			
β	16$_4$	*		
γ	10	26$_{14}$	*	
δ	6	10	16$_4$	*
	α	β	γ	δ

Figure 12.12. Comparison of methods on a matrix of artificial data of 12 characters with three different character-state distributions among four taxa repeated as shown and converted to a matrix of pairwise distances. Topologies and derived matrices obtained from these data are discussed in the text.

designed an example in which the F/M method produces the counterintuitive result of separating the two pairs whose distances give the largest sum of three possible pairs.

The best tree by the Wagner distance procedure separates the taxa into (α, δ) and (β, γ). This is shown in the center-left of figure 12.12, where the subscript values in the matrix show the excess of the phyletic over the observed distance. The Wagner distance procedure will always fit five of six distances perfectly and have a length identical to that of a true minimal-length tree obtained by linear programming, such as the one shown to the right.

The best tree by the parsimony procedure separates the taxa into (α, γ) and (β, δ). This is because there will be 24 character-state changes for the Type II and III characters regardless of tree topology. Thus, the most parsimonious tree is determined by the two Type I characters, which will require only one change of state each on the most parsimonious tree but two each on the other two topologies. Of the three possible topologies, we get a different one of them as best by the three different methods.

There is a fourth method, single linkage, shown at the bottom of figure 12.12, but it is not considered very satisfactory for phylogenetic purposes. One reason is that the taxa are not all at the tips of the tree, although there is an algorithm to answer that topological problem (Fitch 1977) and it gives, in this case, the topology of the most parsimonious tree.

Figure 12.13 is similar but shows the same data analyzed by three methods for two topologies, upper and lower on the figure. On the left are nine characters for four taxa. This example is particularly interesting in that the data are additive, and hence both the F/M method and the linear programming (or Wagner distance) method find the tree which permits the precise reconstruction of the original distances as shown by the two lower trees (left and center) and the lower left half of the phyletic distance matrix immediately above them.

The alternative topology is shown at the top and its phyletic distances shown in the upper right half of the distance matrix in its column. The upper trees are worse by both the length criterion and %SD.

On the far right, the two topologies are examined parsimoniously and it is the upper tree that is most parsimonious (a result determined solely by the first character). We thus have the curious result that *the data are additive for a tree that is not most parsimonious,* yet it is exactly when the data are additive that parsimony and matrix methods ought to give the same answer. Note also that, if one believes that the most parsimonious tree is the best estimate of topology, discarding an F/M tree because it has a negative leg would lead to discarding the best tree, although in this case it would be discarded anyway because of its greater %SD.

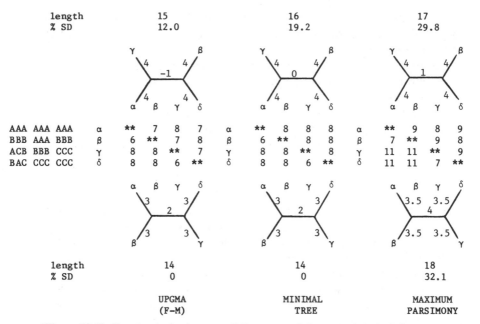

Figure 12.13. Two topologies (upper and lower sets of three trees) derived from a matrix of nine characters for four taxa by three methods.

Table 12.4 shows four taxa with three character types of multiplicity 5, 3, and 1. All methods find the same tree, in which the taxa are separated into (α, β) and (γ, δ). The length of the tree is 11, 12, 13, and 14 by the F/M, minimal-length, maximum-parsimony, and single-linkage methods. The lengths of the terminal legs are two for all legs by all methods, only the interior branch differing in length at 3, 4, 5, and 6 by the same four methods, respectively. (This assumes one does not know the primitive state and the twos assigned are the average of all most parsimonious solutions. If A is primitive, there are still eight changes on the terminal branches, but their distribution is no longer two on each. In single-linkage, the four changes on the two terminal links each represent two terminal phylogenetic branches with two changes each on them.) This result is illustrative of a general property. The length of a tree through the fitting of such data (Fitch and Smith 1982) is necessarily of the form

$$\text{F/M} \leq \text{minimal length} \leq \text{maximum parsimony.}$$

Since the length of the maximum-parsimony tree is not less than the historically correct number of character-state changes, its estimates of the

Table 12.4. Hypothetical data set of 9 characters for 4 taxa
with a taxon-by-taxon distance matrix for these data

===

Character States			Taxon	Taxon			
				alpha	beta	gamma	delta
A A A A A	A A A	A	alpha	–			
A A A A A	C C C	C	beta	4	–		
C C C C C	A A A	C	gamma	6	8	–	
C C C C C	C C C	A	delta	8	6	4	–

----

amount of change must be closer to the truth than those of the methods to its left. The single-linkage estimate is not guaranteed to be greater than the true amount of change, nor even greater than the F/M value.

The final figure using artificial data is figure 12.14 and shows a matrix of pairwise distances (upper right half) that are "additive" in a very loose sense. The phyletic distances on the tree above precisely equal the observed distances in the matrix, but the meaning is loose in the sense that one cannot have a negative distance in "real" life. The Wagner distance procedure necessarily fits five of the six distances but does so on a different tree topology (the subscript 4 shows the only difference between the phyletic and observational distances). In some intuitive sense, the upper tree's topology seems right because alpha and beta are 20 units apart, gamma is 30 or more units from both of them, and delta is 42 more units from all three previous taxa. Again there is no "true" answer. The intuitive appeal of the upper topology is possibly based upon some sense of relatively uniform rates. To the extent that some characters might possess such relative rate properties, a procedure appropriate for them would be useful, assuming such characters were recognizable.

Discussion

I would like to encourage you to adopt a spirit of discriminating eclecticism. The field of systematics has been in considerable turmoil as various investigators developed different methods of classification and argued their merits. I guarantee you that no one method or view has all the good points.

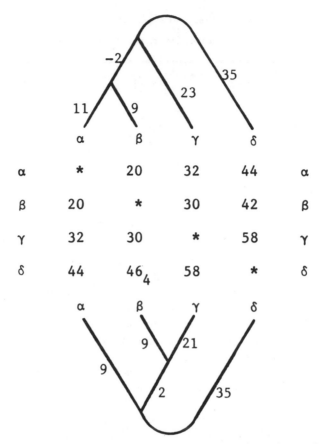

	α	β	γ	δ	
α	*	20	32	44	α
β	20	*	30	42	β
γ	32	30	*	58	γ
δ	44	46₄	58	*	δ

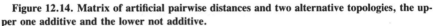

Figure 12.14. Matrix of artificial pairwise distances and two alternative topologies, the upper one additive and the lower not additive.

That is why I commend eclecticism to you; grab the benefit regardless of the label attached. I guarantee you that no method or view is without its blemishes. That is why I suggest that your eclecticism be discriminating.

I would like to conclude by giving you a brief idea of how my attempts at a discriminating eclecticism view the material in this volume. A part of that material is of the how-it-is-done variety. I will concentrate instead on that part which has been of the what's-happening-now-in-research variety.

One of the most interesting papers in this series is by Baum (this volume), who presents us with a data set for *Avena* cultivars whose evolutionary history was completely known from breeding records. Those data were analyzed by most of the methods generally available, and none gave the correct phylogeny. The more striking observation, however, is that *none of the characters seems to fit on the true tree without homoplasy*. And that is over only above 100 years of breeding. Could any observation be more pertinent in telling us that in all our studies of tree-building techniques we have minimized too much our study of characters and their changing states in nature? Your own experience in the laboratory may have been of the type, "My maximal clique has only three characters in it!" Doesn't that say the same thing?

One of the important observations of recent times is that of Felsenstein (1979), who showed, starting from a generalized maximum likelihood expression, that different special conditions led naturally to the legitimacy of some of our tree-building methods. The question then is not whether one should use parsimony or maximal cliques or UPGMA, but how to decide which technique is best to analyze a given character. The presentation by Felsenstein (1979) shows that progress is being made along these very lines.

As Felsenstein has a maximum-likelihood program, I asked him to analyze the data in figure 12.13. My first surprise came when I was told it could not really be done without a knowledge of the number of characters that were identical over all taxa. This arises because the method depends upon an estimate of rate of character-state change that is larger or smaller depending upon whether many or few of the characters have changed their state. Maximum parsimony can, for purposes of discovering the best topology, ignore even characters that change if they change only to uniquely derived states (autapomorphies), but maximum likelihood demands the inclusion of even the unchanging characters.

Because the number of unchanged characters is important, several values were tried. When the number of unchanged characters was zero, the F/M or UPGMA topology was obtained. When the number was one, the topology making α, δ and β, γ the two pairs of sister groups was obtained. When the number was two or greater, the maximum parsimony topology was obtained. Thus, all three possible topologies were obtained using different numbers of unvaried characters with the F/M topology for the fastest rate of character change and the maximum parsimony topology for the slowest rate. This too suggests that the appropriate method is a function of the characters being examined. Smith (1983) seems also to have stumbled upon this idea.

While I demonstrated that the most parsimonious tree for all the characters may not be consistent with that for the maximal clique, a maximal-clique analysis (Estabrook, this volume) might still be the appropriate method under some conditions, as Felsenstein (1979) has shown. I was more taken by Meacham's presentation (this volume) for the following reason.

One can obtain the number of characters with which any given character is incompatible. The more characters with which a given character is incompatible, the less likely it is to be consistent with the most parsimonious tree. Thus, one unbiased, a priori method of weighting is to weight a character in inverse proportion to the number of characters with which it is incompatible. Some refinement would be in order, but the principle is clear.

Little attention has been paid to the polymorphism parsimony approach introduced by Farris (1976) and Felsenstein (1979). More will be paid to it in the future, because it surely represents the truth sometimes. One such possible time is shown in figure 12.15, which shows the apparent monomorphic threonine condition in gamma-135 hemoglobin of man, chimpanzee, and gorilla and the similar alanine in a rhesus and marmoset. Given

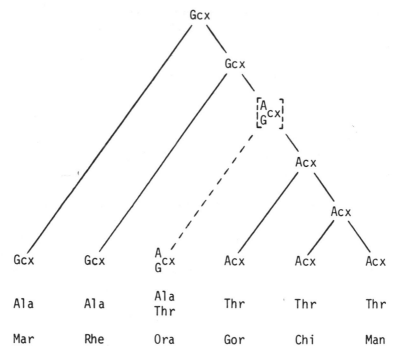

Figure 12.15. Cladogram showing distribution of threonine (Thr) and alanine (Ala) amino acids of gamma-135 hemoglobin in man, chimpanzee (Chi), gorilla (Gor), orangutan (Ora), Rhesus monkey (Rhe), and marmoset (Mar).

the presumed phylogeny and the codons for those amino acids, the most parsimonious evolutionary history is shown by the solid lineages (ignoring the orang). Given the nature of the process, there must have been a time in the lineage descending to the common ancestor of the great apes when the population contained both alanine and threonine and the codon had a mixture of guanine and adenine in the first nucleotide position. Parsimony procedures seldom require an explicit recognition of this intermediate state, but the addition of the orangutan data alters the case, since the orangutan population contains this very polymorphism. Now parsimony requires the insertion of the polymorphism, as shown in the dotted brackets. This leads to the natural inference that it was precisely the polymorphic ancestral population that speciated to give rise to the orangutan with its preserved polymorphism on the one hand, and the higher apes which lost the alanine allele on the other. This too shows that the method needs to be suited to the character and that characters need somehow to be discriminated among themselves for whatever it is that would allow us to assign them to the most appropriate method for their analysis.

In summary, I see the future as providing us with methods having phylogenetic resolving power far greater than those currently available. It will do so by understanding characters better and using methods appropriate to that understanding. In the meantime, take all absolutist pronouncements with a grain of salt, separate scientific utility from the ideological chaff, keep your options open, and enjoy the search for better ways.

Literature Cited

Farris, J. S. 1972. Estimating phylogenetic trees from distance matrices. *Amer. Natur.* 106:645–668.

———. 1978. Inferring phylogenetic trees from chromosome inversion data. *Syst. Zool.* 26:275–284.

Felsenstein, J. 1973. Maximum likelihood and minimum step methods for estimating evolutionary trees from data on discrete characters. *Syst. Zool.* 22:240–249.

———. 1979. Alternative methods of phylogenetic inference and their interrelationship. *Syst. Zool.* 29:49–249.

Fitch, W. M. 1971. Toward defining the course of evolution: minimum change for a specific tree topology. *Syst. Zool.* 20:406–416.

———. 1977. On the problem of discovering the most parsimonious tree. *Amer. Natur.* 111:223–257.

———. 1982. A non-sequential method for constructing a hierarchical classification. *J. Mol. Evol.* 18:30–37.

Fitch, W. M. and E. Margoliash. 1967. The construction of phylogenetic trees—a generally applicable method utilizing estimates of the mutation distance obtained from cytochome *c* sequences. *Science* 155:279–284.

Fitch, W. M. and T. F. Smith. 1982. Implications of minimal length trees. *Syst. Zool.* 31:68–75.

Goodman, M., J. Czelusnick, and G. W. Moore. 1979. Further remarks on the parameter of gene duplication and expression events in parsimony reconstructions. *Syst. Zool.* 28:379–385.

Goodman, M., J. Czelusnick, G. W. Moore, A. E. Romero-Herrera, and G. Matsuda. 1979. Fitting the gene lineage into its species lineage, a parsimony strategy illustrated by cladograms constructed from globin sequences. *Syst. Zool.* 28:132–163.

Hendy, M. D. 1983. On Fitch and Smith's conjecture for minimal length trees. *Syst. Zool.* 32:276–277.

McMorris, F. R. and D. A. Neumann. 1983. Additional comments on Fitch and Smith's

Smith, T. F. 1983. Clustering at the limits. *The Biologist,* 65: in press.

Sarich, V. 1969. Pinniped origins and the rate of evolution of carnivore albumins. *Syst. Zool.* 18:286–295.

Smith, T. F. 1983. Clustering at the limits. *The Biologist,* 65:in press.

Wake, D. B. 1981. Allozymes and phylogeny. Paper presented at Workshop on the Theory and Application of Cladistic Methodology, University of California, Berkeley.

PART IV

Applications

INTRODUCTION

The two principal applications of cladistic analysis are classification and biogeography. Some workers include in the definition of cladistics the construction of a classification directly from the cladogram (e.g., Nelson 1973; Cracraft 1974; Wiley 1981). An alternative view (e.g., Mayr 1981; Felsenstein, this volume) is that the construction of a classification is a separate application. Cladograms are useful explicit statements about the branching patterns of phylogeny, but cladistics does not necessitate their direct use in classification. Most workers, however, would probably be inclined to use the cladogram in some fashion (either as an absolute framework or as a partial guide) for the construction of a classification. The paper by Phillips presented here includes a discussion of the stability of classifications derived from phenetic and cladistic approaches and offers guidelines for the use of cladograms in classification.

The application of cladograms to biogeography has led to a new emphasis: vicariance biogeography. This approach seeks recurring patterns of distributions of organisms in relation to concurrent cladistic patterns (Cracraft 1975; Wiley 1980, 1981; Nelson and Platnick 1981). Such overlapping data suggest generalized "tracks" of phylogeny in time and space that may be explainable by major earth changes (e.g., continental drift, mountain building, etc.; see Croizat 1962; Croizat, Nelson, and Rosen 1974). Advocates of vicariance approaches to biogeography regard these as the only testable (and therefore admissible) biogeographic hypotheses. Many other workers, however, while recognizing the value of major earth events in explaining distributions of organisms, also believe dispersal hypotheses to be a valid approach (Craw 1978; Carlquist 1981). Nelson outlines a method for the examination of distribution patterns in relation to reconstructions of evolutionary history.

Literature Cited

Carlquist, S. 1981. Chance dispersal. *Amer. Sci.* 69:509–516.
Cracraft, J. 1974. Phylogenetic models and classification. *Syst. Zool.* 23:71–90.

——. 1975. Historical biogeography and earth history: Perspectives for a future synthesis. *Ann. Missouri Bot. Gard.* 62:227–250.

Craw, R. C. 1978. Two biogeographical frameworks: Implications for the biogeography of New Zealand. A review. *Tuatara* 23:81–114.

Croizat, L. 1962. *Space, Time, and Form: The Biological Synthesis.* Caracas, Venezuela: privately published.

Croizat, L., G. Nelson, and D. Rosen. 1974. Centers of origin and related concepts. *Syst. Zool.* 23:265–287.

Mayr, E. 1981. Biological classification: Toward a synthesis of opposing methodologies. *Science* 214:510–516.

Nelson, G. 1973. Classification as an expression of phylogenetic relationships. *Syst. Zool.* 22:344–359.

Nelson, G. and N. Platnick. 1981. *Systematics and Biogeography: Cladistics and Vicariance.* New York: Columbia University Press.

Wiley, E. O. 1980. Phylogenetic systematics and vicariance biogeography. *Syst. Bot.* 5:194–220.

——. 1981. *Phylogenetics: The Theory and Practice of Phylogenetic Systematics.* New York: Wiley.

13

Considerations in Formalizing a Classification

RAYMOND B. PHILLIPS

Classification, followed by the application of names, is an essential part of human thought and communication. It is an activity that summarizes observed patterns. The best classifications are those that are based on the largest possible pool of information evaluated by a means appropriate to the use to be made of the classification, thus conveying the largest amount of information.

In botanical classification, the work of such pre-Darwinians as Jussieu (1789) and Candolle (1813) resulted in classifications commonly referred to as "natural" in many introductory plant taxonomy textbooks (e.g., Benson 1979; Jones and Luchsinger 1979). These systems, particularly at the generic and familial ranks, were based on a large assortment of characters observed in the plants, and groups were defined by the observed similarities among them. Most of the changes found in classifications since then at these ranks have been primarily the result of the acquisition of a large amount of new data (see Engler and Prantl 1898; Takhtajan 1969; Cronquist 1981). The impact of evolutionary theory on biological classification has been primarily as a means of explaining the patterns of similarity that are observed and incorporated in the classifications, rather than directly affecting the classifications. Evolutionary theory has had so little impact because the same operational approach has been used throughout most of the recent history of classification. Namely, similar organisms are grouped together, the gaps between groups of organisms, as evidenced by the distribution of character states, providing the basis for the classification. The recognition of such groupings and gaps does not require evolutionary theory, although they may be interpreted in an evolutionary context. One of the strongest and most explicit statements in support of this approach is that of McNeill (1979):

"Classification, on a phenetic basis, and phyletic-tree reconstruction, on a cladistic one, are separate and complementary activities; let us keep them that way."

A significant departure from this approach to classification is evident in the work of Hennig (1966), forming the basis of cladistic methodology. According to Hennig, gaps are relatively unimportant, only genealogical relationships being of value in grouping organisms in a classification serving as a general reference system (Farris 1979).

In many cases grouping either by overall similarity or by genealogical relationship will lead to the same groups, because members of a genealogical lineage may be expected to be more similar to each other than they are to members of another lineage. However, for the particular assortment of characters used to construct a classification this may not always be true. It is toward a resolution of this kind of conflict that I want to work in this brief discussion.

The results of a cladistic analysis provide two kinds of information about the evolutionary history of a group: the branching pattern—cladogenetic change—and the amount of divergence in observed characters along each branch—anagenetic change. For example, in figure 13.1, which represents a cladogram for a hypothetical group consisting of four taxa, the estimated rate of anagenetic change (indicated by the number of cross-lines, each of which represents one character-state change) is unequal along the branches. As a result, B is in a sense left behind evolutionarily by the group C and D, on the basis of the characters observed. Consequently, B is more similar to A than it is to its own sister group, the lineage leading to C and D. There is a gap. A classification may clearly reflect either the gap or the genealogical relationships between the organisms. It cannot fully reflect both kinds of information. Therefore a choice must be made as to which is more important or useful. A classification that attempts to summarize both at the

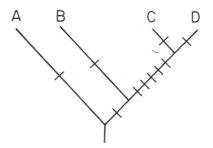

Figure 13.1. Cladogram of a hypothetical group of four taxa. Cross-lines on branches indicate synapomorphies of the taxa to which the branches lead.

same time in a conventional Linnaean hierarchy will be ambiguous while trying to be balanced (Farris 1976). Only by reference to the taxonomist's accompanying narrative explaining the basis for group recognition will the suggested relationships be clarified. The basic question, therefore, appears to be to what degree a classification alone can be used to convey information about the organisms, besides serving as an "index" to the literature.

The Purposes of a Classification

Probably most modern biological systematists would agree that the primary purposes of a classification are to summarize some aspect of evolutionary relationships and, thereby, to be informative about the distribution of character states (i.e., information content should be high). A classification that does this best is to be preferred over alternatives. However, there is considerable disagreement over what is meant by evolutionary relationship and over the means of comparing classifications as to their information content.

Evolutionary relationship is usually defined as either genetic similarity or common ancestry (genealogy; Wiley 1981:261). The view that there is not necessarily concordance between these definitions has been advocated by Mayr (1969, 1974, 1981), Bock (1974), and Ashlock (1971). However, Wiley (1981:262–263), after citing examples of perceived discordance by Mayr and studies of molecular genetic similarities in several groups of vertebrates, concludes that "genetic evidence is concordant with the postulated genealogical relationships not with the families recognized by Mayr. Further, these works demonstrate that overall phenetic similarity can be a poor estimator of genetic similarity." In these cases, and perhaps in general, it appears to me that genealogical relationships, and classifications based thereon, are the best predictors of overall genetic similarity, though not necessarily of more restricted aspects of genetic similarity, such as external morphological similarity.

The problem of comparing the information content of classifications has, in recent years, resulted in heated debate. One approach is to compare the correlation between the classifications and the similarity or distance matrices from which they are constructed, using the cophenetic correlation coefficient as a measure of the preservation of information by the classification (Sneath and Sokal 1973). Farris (1979), Mickevich (1978), and Mickevich and Johnson (1976) have found consistently higher correlation coefficients for cladistic classifications than for those based on overall similarity by cluster analysis. Also involved in these studies is the congruence between classifications based on different data sets for the same organisms when various methods are applied. It is concluded that cladistic classifications appear superior in their stability. However, Rohlf and Sokal (1980, 1981) have criticized these results and, through alternative methods of analysis, reached the conclusion that neither approach is clearly superior at preserving information content or in stability.

There appear to be two difficulties involved in this kind of comparison. First, the use of a cophenetic correlation to assess the ability of different classifications to preserve information should be with reference to the same data matrix, not to different matrices, in order to provide a useful comparison (Rohlf and Sokal 1980, 1981). Second, the phenetic classification to which a cladistic classification is compared is the phenogram. However, the phenogram in practice is never entirely reflected in a phenetic classification, although it could be. Only some of the groups are recognized formally in the Linnaean hierarchy (Sneath and Sokal 1973), generally those most distinct. Consequently, considerable additional information is lost, while achieving as simple a classification as possible to reflect major patterns.

A second approach to evaluating the information content of a classification is by the use of information theoretic optimality criteria. Duncan and Estabrook (1976) have used a sum-of-fractions statistic (sum over all characters of the redundancy between each character and a classification, divided by the information in the classification) as the basis for evaluating the groupings at a particular rank for efficiency in preserving the character information. If, for instance, we wish to compare classifications at an intermediate rank for the taxa in figure 13.1, two groupings might be proposed. The first is the highest rank within the group based on the cladistic history, namely A and BCD. The second is based on overall similarity, namely AB and CD. The sum-of-fractions statistics for these classifications at this particular rank are 4.37 and 6.56, respectively. At this rank the classification based on overall similarity is much more efficient at preserving character information. However, this is not a completely valid comparison, because the cladistic classification would contain another rank, or at least a convention for describing more detailed relationships, as discussed below. In order for the comparison to be meaningful, the entire cladistic classification for the group must be compared with the entire classification based on overall similarity. The method of Duncan and Estabrook is not able to make such a comparison.

Another way of determining the information content of a classification is by examining the number of types of character-state distribution predicted by the classification (Farris 1979). A character type is one or more characters with a particular distribution of character states. Any character with this distribution of character states belongs to this character type. There are two types of characters in figure 13.1, excluding the four autapomorphies that are of no value in determining genealogical affinity. The first type, having one state in A and the other in B, C and D, by its synapomorphic state defines the group BCD as monophyletic and consists of one character. The second type, having one state in A and B and the other state in C and D, by its synapomorphic state defines the group CD as monophyletic and consists of five characters. A cladistic classification contains information that predicts both types of characters, while one based on overall similarity predicts only the characters of the second type. Furthermore, the latter classification does not predict which group (AB or CD) has the apomorphic state for characters belonging to the second type. It does not tell us which group has the plesiomorphic state and, hence, the distribution of the character states in other organisms. A classification based on overall similarity allows prediction of the most frequently occurring type(s) of character-state distribution but predicts little or nothing about other types and nothing about the distribution of the character states outside of the group being considered. On the other hand, a cladistic classification far more accurately describes the types of character-state distribution, both within the group and in other organisms, but does not predict which is the most frequently occurring type of distribution for the assortment of characters observed in a particular study. It would seem that, since the frequency of character types may be highly dependent on the character sample used in a study, a classification based on overall similarity would be less stable upon addition of new characters or use of alternative characters. This is probably the reason, at least in part, for some findings of higher congruence between results of cladistic analyses than for those of phenetic analyses when different characters are used in studies of real organisms. In essence, gaps may come and gaps may go, depending almost entirely on the particular sample of characters used in a study (Mickevich 1978), because "the rate[s] of divergence of different characters are often distinctly different" (Mayr 1981).

There is, however, another aspect of classificatory stability that must be considered. An effect on cladistic results is observed when hypotheses about character-state polarity are altered. Figure 13.2 represents such a cladogram derived from the same data used in figure 13.1 after reversal of polarities for characters of the two character types described above. A cladistic classification will be substantially altered by these changes, while one based on overall similarity will not be affected, because hypotheses about polarity are not included in the latter analysis.

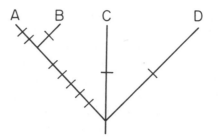

Figure 13.2. Cladogram for the hypothetical group of taxa in figure 13.1 after reversal of character state polarities for six characters.

Paraphyletic Groups

In spite of their failure to represent genealogical (and thereby genetic) relationships, classifications that contain one or more paraphyletic groups (such as group AB in figure 13.1) continue to be advocated under some circumstances when cladistic methods are applied. Duncan (1980) has employed a convexity criterion (Estabrook 1978) for determining whether a classification is consistent (sensu Simpson 1961) with a proposed phylogeny. A taxon is convex if the phyletic line segments joining members of that taxon do not intersect the phyletic line segments joining members of any other taxon at the same rank. An application of this criterion is illustrated in figure 13.3, in which two classifications are examined. In the first classification (figure 13.3a), which consists of two taxa (AB and CD), both taxa are convex, because there is no intersection between the two sets of phyletic line segments. According to this criterion, the classification is acceptable, although taxon AB is paraphyletic and taxon CD is monophyletic. The second classification (AD and BC) is unacceptable because it is inconsistent with the phylogeny, since neither taxon is convex. This is because the most recent common ancestor of C and D must belong to both taxa, violating the no-overlap requirement of conventional classification or requiring the independent evolution of members of one of the taxa (polyphyly). The purpose of the convexity criterion's application to classification is, therefore, to exclude polyphyletic taxa. There are, however, several classifications for these four objects that satisfy the convexity criterion. The choice from among them is determined by a synthesis of a variety of lines of evidence, as has been done traditionally (Duncan 1980). This approach has been called eclectic (Duncan 1980; Mayr 1981) or syncretistic (Farris 1979).

The use of the convexity criterion allows the taxonomist considerable

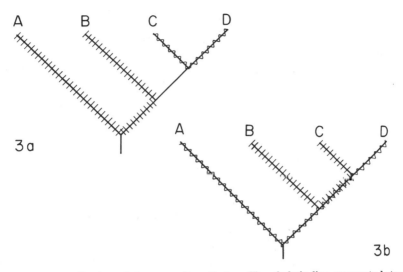

**Figure 13.3. Application of the convexity criterion. The phyletic line segments between
members of a class are marked for two classifications: (a) taxon 1: A and B; taxon 2: C and
D, (b) taxon 1: A and D; taxon 2: B and C. See text for explanation.**

freedom in choosing among several possible classifications (Estabrook 1981).
One of these classifications will be a cladistic one, in which only monophyletic taxa are recognized and in which the entire genealogical history
of the group is reflected. As noted by Estabrook (1978, 1981) and Duncan
(1980), the recognition of only monophyletic taxa is a special case of the
application of the convexity criterion, and a very restrictive approach to
selecting the classification.

There are some interesting aspects to the application of the convexity
criterion when paraphyletic groups result. The most recent common ancestor of B, C, and D in figure 13.3a must be a member of taxon AB. However, through a speciation event it has given rise to the lineage that leads
to the ancestor of taxon CD. If the rank concerned in this classification is
genus, it could be said that a member of one genus gave rise to another
genus through a speciation event, or that the ancestor of one genus is another genus. The difficulties with this concept have been discussed at length
by Wiley (1981), who concludes that higher taxa cannot give rise to other
higher taxa. However, his position is partly impractical, because ancestors
of groups above the species level could not be incorporated in the classification. For example, if the most recent common ancestor of two genera
is identified in the fossil record, how should it be named? Wiley's convention eight (1981:223) requires that in such cases the ancestor be classified

in a monotypic genus and listed in parentheses beside the taxon containing
its descendant. It certainly cannot be a member of either genus, or a paraphyletic group would be established. For the same reason the ancestor cannot
be a member of any genus that contains another species—again a paraphyletic group would be formed. Therefore, the genus containing the ancestor
must be monotypic. In that case, however, we have a member of one genus giving rise to two new genera through a speciation event, again an un-
satisfactory arrangement. The only solution is to consider both of the genera and the ancestor to be members of the same genus. Unfortunately, this
solution would continue through the entire phylogenetic tree of life on earth,
requiring all living things to be members of the same genus. The problem
is caused by the requirement of our Linnaean system that all taxa be members of taxa at certain ranks, making the classification of fossil organisms
difficult when only monophyletic taxa are allowed.

The recognition of paraphyletic taxa remains an attractive option for many
taxonomists. This attraction is simply a consequence of the intuitively observed patterns of overall similarity, and the option is one most taxonomists
resist abandoning, partly because of one common way of viewing the nature of character-state distributions. Referring again to figure 13.1, most
taxonomists would probably recognize two taxa (AB and CD) in the absence of a cladistic analysis, because B is more similar to A than it is to C
or D on the basis of overall similarity with the characters currently available. However, what is the nature of this similarity? For the characters considered here, the only character states that A and B share are plesiomor-
phies, so that the paraphyletic taxon AB is based on shared ancestral
character states (symplesiomorphies). If other organisms had also been included in an analysis of overall similarity for the group, many of them, in
particular the out-group, would also share the plesiomorphic state and
consequently have to be included in the taxon AB. Having a plesiomorphic
binary character state means that nowhere in the evolutionary history of
that taxon has the modification resulting in the apomorphic state occurred
(excluding homoplasy, with which I am not now concerned). Consequently, all other living and extinct organisms would have to be included
in taxon AB except C and D, since none of them have in their evolutionary
history the modification that appeared in the most recent common ancestor
of C and D. Of course this does not happen in practice, because when
overall similarity is evaluated only a restricted number of organisms are involved. Usually a study involves a monophyletic group and as such concerns a collection of organisms that have at least one synapomorphy. If there
is one synapomorphy for the group ABCD in figure 13.1, what is the biological significance of the taxon AB? A and B share the apomorphic state

of the character changing at the base of that tree, but they also share it with C and D. A and B share seven plesiomorphic states, but they also share them with all other living and extinct organisms on earth that are not shown in the cladogram. The only unique feature of taxon AB is its combination of apomorphic and plesiomorphic character states. In other words, taxon AB is the intersection of two sets: the group ABCD, defined by at least one synapomorphy; and the group AB and all living and extinct organisms not included in the cladogram, defined by the five plesiomorphic states which change to the apomorphic state along the phyletic line segment leading to the most recent common ancestor of C and D (in other words, everything but the lineage leading to C and D).

The particular combination of apomorphic and plesiomorphic character states in a paraphyletic group does not come about by a single speciation event, as does a monophyletic group, but rather by two or more such events. In the example shown in figure 13.1, a speciation event gave rise to the lineage leading to the group illustrated. A second speciation event gave rise to the lineage leading to the ancestor of C and D. The paraphyletic group AB is defined with reference to these two events. Consequently, it could be said that paraphyletic groups do not evolve. Instead they are what is left over when another group evolves away from them. An absence of this kind of anagenetic change (or rather the lack of evidence indicating such a change) along a lineage leading away from such a paraphyletic group would cause that group not to be circumscribed in the same way. Paraphyletic groups are defined not by unique evolutionary events but by reference to features restricted to them or shared with other groups. Monophyletic groups are, on the other hand, defined by unique evolutionary events.

Paraphyletic groups are convenient, however, for describing one kind of character state, namely, an intermediate in a transformation series, or character-state tree. For example, if a character has three states and the hypothesized polarity is

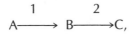

$$A \xrightarrow{\ 1\ } B \xrightarrow{\ 2\ } C,$$

three taxa may be easily recognized by similarities on the basis of this character. Such a classification could be considered natural in that it reflects one aspect of evolution and the taxa would be convex. However, only the taxon whose members have state C would be monophyletic, the other two being paraphyletic. For example, the character might represent ovary position, in which state A is superior, B half inferior, and C inferior, or the character might represent aspects of chemical evolution in which two new enzymes successively appear in the evolution of the group.

There is no question that only a paraphyletic group can adequately describe the distribution of these character states. Unfortunately, a great deal of information is lost by treating a group in this way. A classification consisting of three taxa, each circumscribed by the presence of one of the states, will have no information about the transition series (i.e., which taxon has the apomorphic and which the plesiomorphic state). Neither will it contain any information about other characters, except those that have precisely the same state distribution or the synapomorphic state coinciding with the distribution of state C for binary characters. This is a special classification and not one that is appropriate as a general reference system (Farris 1979; Wiley 1981).

A more appropriate approach is to view such a transition series as several binary characters, each representing the polarity about a single evolutionary event. The above three-state character would then be redescribed using two binary characters. The first would have the apomorphic state in all taxa having either B or C. That is, these taxa have the first event in their evolutionary history. The second binary character would have the apomorphic state only in those taxa having C. Those taxa having state B are apomorphic with respect to the first of these evolutionary events but plesiomorphic with respect to the second.

A classification by itself, of course, contains no specific information about the distribution of character states for any specific character. It does however, allow the user to make certain kinds of predictions based on the degree of relationship. If a derived character state is observed in a particular species, are we more likely to find that character state in other members of a paraphyletic group or in those of a monophyletic group in which that species is classified? If the ancestor of that species had that character state, then all descendants of that ancestor will have the same state or a derivation of that state. The likelihood that a further modification of the derived state will occur along any particular lineage descended from that ancestor is dependent on the overall rate of evolution. For any particular character, however, the likelihood is probably not large. For the group illustrated in figure 13.1, if a derived character state is found in B, is it more likely that the state will also be found in A or that it occurs in C and D? If it is not an autapomorphic state, in which case no other species will have it unless there is parallel evolution somewhere, then an ancestor in common with at least some of the other groups must have had the state. If the state first arose prior to the most recent common ancestor of B, C, and D, then C and D will also have the state. The recognition of the monophyletic group BCD allows prediction of the distribution of the state. The only circumstance, other than parallel evolution, in which the state found in B would occur in

A but not in C and D would be if the most recent common ancestor of A, B, C, and D had the state and there were a change in the state along the lineage leading to the most recent common ancestor of C and D. Unless the probability of this second change is large (greater than 50 percent) we are not likely to find that state restricted to A and B. That any particular character has a 50 percent or greater likelihood of changing along a particular branch in the phylogenetic tree is extremely improbable. Consequently, a classification based on overall similarity in which paraphyletic groups are recognized would have a small chance of allowing the user to search efficiently among other species for additional occurrences of an advanced character state found in one species.

One final comment regarding the problems with recognizing paraphyletic groups needs to be made. Paraphyletic groups are of interest because they have unique associations of particular apomorphic and plesiomorphic states. However, only a very few of the many possible associations are readily apparent in such a classification. In addition to showing genealogical and, by inference, genetic relationships, a cladistic classification allows the prediction of all the possible types of character-state associations. For example, in figure 13.1, a cladistic classification also allows us to predict that A and B share certain plesiomorphic and apomorphic features. Furthermore, BC, AC, ABC, ACD, etc. are all groups of species that have unique associations of plesiomorphic and apomorphic states, which will be evident only from a cladistic classification or reference to a cladogram. This leads me to the conclusion that a cladistic classification, when it can be obtained, in addition to incorporating virtually all the information in a classification based on overall similarity, provides a far more useful general system of reference.

Classificatory Conventions

Two properties of classification are required by many advocates of a phylogenetic (or cladistic) classification. First, all taxa recognized in a classification must be monophyletic (sensu Hennig 1966), and second, the classification must reflect entirely the inferred cladistic history of the group under consideration (Farris 1979; Nelson 1974; Wiley 1981). If some of the cladistic history is left out of the classification, then information is lost. In other words, given a classification it should be possible to reconstruct the inferred genealogical relationships among all members of the group, because there is a one-to-one correspondence between that classification and the relationships inferred.

Critics of the cladistic approach to classification claim that a classification reflecting the group's entire cladistic history will often be excessively complex, requiring many intermediate ranks, thereby decreasing its convenience or utility to the general scientific community (e.g., Gingerich 1980). This criticism is probably well founded if the cladistic history is reflected by requiring the recognition of sister groups at a coordinate rank, as advocated by Hennig (1966). This convention would require, for the example in figure 13.1, where A, B, C, and D are species in a genus, a classification of the following form:

```
Genus 1
    Subgenus 1
        species A
    Subgenus 2
        Section 1
        species B
        Section 2
        species C
        species D
```

Two intermediate ranks and four names are required between genus and species in this example.

This is not necessary, however, if a sequencing convention is used (Nelson 1972, 1974; Wiley 1979, 1981). According to this convention, "asymmetrical trees containing a number of monophyletic groups could be placed at the same categorical rank and listed in order of their branching sequence" (Wiley 1981:206). Application of this convention to figure 13.1 would produce the following classification:

```
Genus 1
    species A
    species B
    species C
    species D
```

Given this classification, and knowing that the sequencing convention was used, we would be able to reconstruct the inferred genealogical relationship in the group. A variety of additional conventions have been developed by Wiley (1981) for dealing with multifurcations (by designating the monophyletic taxa *sedis mutabilis*), taxa of uncertain position *(incertae sedis),* fossil groups including ancestors of extant groups, and hybrids, as well as the inclusion of biogeographic information and degree of anagenetic change.

The annotated system of classification advocated by Wiley (1981) is easily applied once the conventions are understood and need be no more complex than classifications not reflecting genealogical relationships. A tremendous amount of inferred information about genealogies can be stored in a classification by the systematist constructing it, and the information can be readily retrieved by others interested in the group.

Some Practical Difficulties and Possible Solutions

In the hypothetical examples discussed above the characters were presumed to have undergone strictly divergent evolution. The construction of a cladogram in such cases is straightforward, no matter how many taxa or characters are used. However, in most groups studied at least some homoplasy has occurred in the characters used. Sometimes the amount of homoplasy is quite high. In a cladogram of selected families of flowering plants (Young 1981), I have determined that over 300 character-state changes occur; the minimum number of changes required if only divergent evolution had occurred is 58. Each binary factor (there being several three-state characters), on the average, changes about five times.

The problem that homoplasy presents in reconstructing cladistic relationships because of methodological differences in the likelihood of parallelisms and reversals has been discussed by Felsenstein (1978, 1979, and this volume). For a particular set of data and hypotheses about character-state polarity, each different method may produce a different estimate of the cladistic history. The only sure way to obtain only a single cladogram on which to base a classification is to do only one analysis. The reliability of any particular method for determining the cladistic history of a group will depend on the particular characters that are used and the degree to which the method's assumptions about likelihood are satisfied. The dependence of results on methodology in some cases is a serious limitation in cladistic analysis (Hull 1979).

In view of the considerable advantages of constructing a cladistic (phylogenetic) classification, but also of the difficulties in obtaining a reliable estimate of the cladistic history in some groups, the following may serve as a useful set of guidelines.

1) When a reliable estimate of the cladistic history of a group can be obtained, a cladistic classification should be formalized. The classification should contain a preliminary statement indicating that it contains only monophyletic taxa and, if necessary, explaining any special conventions used, such as phyletic sequencing. A reliable

estimate may be expected when there is strong evidence for the hypothesized character-state polarities and relatively little homoplasy.

2) If only a part of the cladistic history, such as some main lineages, can be determined with confidence, then taxa should be monophyletic, as far as possible. For example, several clearly monophyletic groups may be evident but the confident resolution of clades at the base of the tree is not possible with the available information. A paraphyletic group may be recognized for these taxa, with appropriate notation in the classification indicating its status, pending additional study.

3) If a cladistic analysis is impossible, because there is little or no basis for hypothesizing character-state polarity (as when an appropriate out-group cannot be identified), taxa should be convex on an unrooted tree. A previous classification whose taxa are convex on the tree should probably be retained unless there are particular reasons why an alternative would be easier to use or more informative. A classification in which taxa are convex will generally undergo relatively little change once a rooted cladogram can be constructed, because at least some of the taxa will be monophyletic. It is far better to have a general classification deficient in some respects than to arbitrarily root an unrooted tree in order to obtain a completely resolved, though unreliable, cladogram and classification that are likely to change dramatically on further study.

Acknowledgments

I would like to express my gratitude to Thomas Duncan and Tod Stuessy for many useful discussions, as well as to members of the University of Oklahoma Systematics Luncheon Group (SLUG), especially Gary Schnell, James Estes, Charles Daghlian, and Charles Harper. Without their helpful comments this paper would not have been possible. This does not mean, of course, that unanimity has been achieved. My thanks also to Beverly Richey for typing numerous revisions. Support from the Oklahoma Biological Survey is gratefully acknowledged.

Literature Cited

Ashlock, P. H. 1971. Monophyly and associated terms. *Syst. Zool.* 29:63–69.

Benson, L. 1979. *Plant Classification.* 2d ed. Lexington, Mass.: Heath.

Bock, W. J. 1974. Philosophical foundations of classical evolutionary taxonomy. *Syst. Zool.* 22:375–392.

Candolle, A. P. de. 1918. *Théorie Élémentaire de la Botanique*. Paris.

Cronquist, A. 1981. *An Integrated System of Classification of Flowering Plants*. New York: Columbia University Press.

Duncan, T. 1980. Cladistics for the practicing taxonomist—An eclectic view. *Syst. Bot.* 5:136–148.

Duncan, T. and G. F. Estabrook. 1976. An operational method for evaluating classifications. *Syst. Bot.* 1:373–382.

Engler, A. and K. Prantl. 1898. *Syllabus der Pflanzenfamilien*. Berlin.

Estabrook, G. F. 1978. Some concepts for the estimation of evolutionary relationships in systematic botany. *Syst. Bot.* 3:146–158.

——. 1981. Review of the Willi Hennig memorial symposium. *Syst. Bot.* 6:95–100.

Farris, J. S. 1976. On the phenetic approach to vertebrate classification. In M. K. Hecht, P. C. Goody, and B. M. Hecht, eds., *Major Patterns in Vertebrate Evolution*, pp. 823–850. New York: Plenum Press.

——. 1979. The information content of the phylogenetic system. *Syst. Zool.* 28:483–519.

Felsenstein, J. 1978. Cases in which parsimony or compatibility methods will be positively misleading. *Syst. Zool.* 27:401–410.

——. 1979. Alternative methods of phylogenetic inference and their interrelationship. *Syst. Zool.* 28:49–62.

Gingerich, P. D. 1980. Paleontology, phylogeny, and classification: An example from the mammalian fossil record. *Syst. Zool.* 28:451–464.

Hennig, W. 1966. *Phylogenetic Systematics*. D. D. Davis and R. Zangerl, trans. Rpt. 1979. Urbana: University of Illinois Press.

Hull, D. L. 1979. The limits of cladism. *Syst. Zool.* 28:416–440.

Jones, S. B., Jr., and A. E. Luchsinger. 1979. *Plant Systematics*. New York: McGraw-Hill.

Jussieu, A. L. de. 1789. *Genera Plantarum Secundum Ordines Naturalis Disposita*. Paris.

McNeill, J. 1979. Purposeful phenetics. *Syst. Zool.* 28:465–482.

Mayr, E. 1969. *Principles of Systematic Zoology*. New York: McGraw-Hill.

——. 1974. Cladistic analysis or cladistic classification? *Z. zool. Syst. Evolut.-forsch.* 12:94–128.

——. 1981. Biological classification: Toward a synthesis of opposing methodologies. *Science* 214:510–516.

Mickevich, M. F. 1978. Taxonomic congruence. *Syst. Zool.* 27:143–158.

Mickevich, M. F. and M. S. Johnson. 1976. Congruence between morphological and allozyme data in evolutionary inference and character evolution. *Syst. Zool.* 25:260–270.

Nelson, G. J. 1972. Phylogenetic relationship and classification. *Syst. Zool.* 21:227–231.

——. 1974. Classification as an expression of phylogenetic relationships. *Syst. Zool.* 22:344–359.

Rohlf, F. J. and R. R. Sokal. 1980. Comments on taxonomic congruence. *Syst. Zool.* 29:97–117.

——. 1981. Comparing numerical taxonomic studies, *Syst. Zool.* 30:459–490.

Simpson, G. G. 1961. *Principles of Animal Taxonomy.* New York: Columbia University Press.

Sneath, P. H. A. and R. R. Sokal. 1973. *Numerical Taxonomy.* San Francisco: Freeman.

Takhtajan, A. 1969. *Flowering Plants: Origin and Dispersal.* C. Jeffrey, trans. Washington, D.C.: Smithsonian Institution Press.

Wiley, E. O. 1979. An annotated Linnaean hierarchy, with comments on natural taxa and competing systems. *Syst. Zool.* 28:308–337.

——. 1981. *Phylogenetics: The Theory and Practice of Phylogenetic Systematics.* New York: Wiley.

Young, D. A. 1981. Are the angiosperms primitively vesselless? *Syst. Bot.* 6:313–330.

14

Cladistics and Biogeography

GARETH NELSON

An earnest desire to make cladograms for various groups of organisms can have but one result: as time goes by there will be more, rather than fewer, of these constructions. As they proliferate problems will be created at two levels: the particular and the general. The particular level is that of a group, such as buttercups, and the various cladograms different people might make of the interrelationships of the different species. What do people do with their different cladograms? Often they argue about them in the belief that, despite the diversity of results, one is true, others false. I pass over the notion that, in some sense, all such cladograms are true even if they truly disagree. This notion implies no possibility for progress beyond the mere accumulation of cladograms.

Increase the number of groups and the possibilities for argument are thereby increased. Such has been the nature of systematics for hundreds of years. In the early days a good deal of the argument was submerged in history. At any one time there was apt to be but one person intensely working on any one group. The only possibility for argument was with those who had gone before. If they were already deceased, the living had only to open their mouth to have the last word, but only for the time being. Yet that was enough to create the illusions of authority and final truth in taxonomy. Now times are different, and there are fewer groups that are the province of the lone authority-for-the-time-being. However, to one degree or another the illusions persist.

According to my personal experience cladistics developed because approximately 15 years ago a number of us were working on basically the same problem, and we had each other to argue with. Much argument ensued, but it eventually passed away as a cladistic approach to the problem was realized. In retrospect, I believe that some progress was made; in other words, a particular problem was solved. Since then, I have seen other problems solved in much the same way. Hence, I have reason to believe that the particular level is under some control; that, in general, one cladogram really is true and others false; and that with some effort progress toward the truth may be achieved.

For progress at the general level a common denominator must exist that brings cladograms of different groups into some kind of relation. Evolutionary laws, if there were some, might suffice as a common denominator, but I know of none. What do cladograms of different groups have in common, beyond the theory of systematics that underlies them? Only two factors come to mind: space and time. Thus, biogeography is where our efforts come together.

Immediately a procedural objection arises, if I may term it that. The typical authority-for-the-time-being is unwilling to consider that, for example, the truth that he seeks in buttercups is better displayed by liverworts, or some other group. If he believes he has already found the truth of buttercups, he is likely to be unreceptive to the notion that liverworts might prove him wrong, after all. If he has spent his life with buttercups, he has no time for liverworts at all.

Science has ways of accommodating itself to the limits of human receptiveness and mortality. Within biogeography the accommodation was achieved through the notion of diverse centers of origin: whereas buttercups have one center of origin, liverworts can, and no doubt do, have another; there need be no agreement between the two, and therefore there need be no disagreement among living members of the botanical fraternity. In matters of detail each specialist can go his own way, with the tacit understanding that his right to pose as an authority-for-the-time-being will be unchallenged until after he is dead.

Unless I grossly misread history, all of this is of the past, thanks chiefly to Croizat. Now it seems clear that notions of center of origin have their roots in creation myth—the Garden of Eden—not in empirical science. From this standpoint an alternative has emerged—distributions of diverse groups are but variations on one single theme of earth history. Still, variation from group to group is a problem. As I see it the problem is an empirical one. By this I do not mean that everyone sees the problem as I do. On the contrary, I am well aware of the reputation that Croizat has enjoyed among the botanical fraternity, at least in past years. If reputation is to be considered relevant, someone else will have to argue the case. For the moment I deal with cladograms as viewed in the geographical dimension.

Distribution of taxa are more or less endemic or cosmopolitan. For a variety of reasons endemic taxa offer the best possibilities for an examination of cladistic relationships in a geographical dimension. One reason is easy to understand: if we speak of relationship in a geographical sense, we must have in mind some geographical units that are interrelated. In the cladistic sense, the smallest units of this sort are areas of most local endemism. Thus, area relationship is but the sum of the relationships of endemic taxa.

An example may illustrate the point. Consider a group with four species endemic to four different areas: Japan, the Philippines, Queensland, and New Caledonia. If the species can be placed in an informative cladogram, such as that shown in figure 14.1A, there would be reason to say that the four areas, as evidenced by this particular group of four species, are inter-related in exactly the same way (figure 14.1B). Area relationship in this sense is but the sum of the "facts," as far as they are known, of species relation-ships.

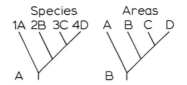

Figure 14.1. A. Cladogram of four species (1–4) in four areas (A–D). B. The correspond-ing area cladogram.

The basic question is whether cladograms for different groups of organ-isms interrelate areas in the same way or ways. At present, there is no easy way to answer definitively this question. Preliminary attempts, and among these I include those of Brundin (1966), Rosen (1979), Patterson (1981), and virtually all other efforts of this sort to date, suggest that, as Croizat has so often argued, distributional information is exceedingly redundant. In other words, geographical agreement among different groups may be the rule rather than the exception. If so, this finding would have some significance, for it is contrary to the expectations of most so-called traditional approaches to biogeography and to those of neo-Darwinism. Thus we may understand why the fraternity feels as it does. I would not suggest that they perceive the fox already in the coop. That suggestion would imply too much perception. At best, they perceive only the vague outline of some alien intrusion raising a fuss.

There are many reasons why the basic question is difficult to answer de-finitively at present. Perhaps the main reason is the scarcity of cladograms, especially for taxa at a low rank, e.g., subspecies of birds or species within genera of flowering plants. Other problems are analytical, and in what fol-lows I address those which concern cladistic structure and congruence. I have time to deal only with a few aspects; for a fuller treatment consult the relevant literature (e.g., Nelson and Platnick 1981) or develop your own analytical procedure.

Consider an ideal case of disagreement, or noncongruence, of two species cladograms (figure 14.2A, B). Under the assumption that the two groups are monophyletic, combining the two as species cladograms simply results in a more complex species cladogram (fig. 14.2C). Under the assumption that they might not be monophyletic, combining them gives an uninformative result (fig. 14.2D). Considering only the area cladograms (fig. 14.2E, F), we may consider two possible results, complex and simple (fig. 14.2G, H), neither of which has anything to do with assumptions of monophyly. Both results show that there is no geographical pattern common to the two groups of species. The complex result (fig. 14.2G) implies that there are two different sets of areas; that, in some sense, A≠A, or B≠B, or C≠C, or D≠D, or some combination of these inequalities. The simple result (fig. 14.2H) is cladistically uninformative; it offers no explanation for the variation and in that sense implies that the variation is random. Likewise, it says nothing about whether there is one or two sets of areas.

Now consider an ideal case of agreement, or congruence, of two species cladograms (fig. 14.3A, B). The same possibilities exist for combining species cladograms (fig. 14.3C, D) and area cladograms (fig. 14.3E–H). In this case the area cladograms are the same, and the simple result (figure 14.3H) is fully informative. The only problem here is to judge whether the agreement is significant or due simply to chance—in other words, whether the evidence of a nonrandom pattern is sufficient to conclude that the pattern

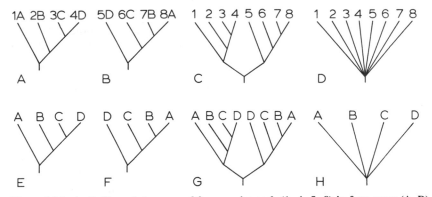

Figure 14.2. A, B. Two cladograms of four species each (1–4, 5–8) in four areas (A–D). C, D. Combination of the two species cladograms under the assumptions that each is a monophyletic group (C) and that each might not be a monophyletic group (D). E, F. Area cladograms derived from A and B. G, H. Possibilities for combination of area cladograms: complex in G and simple in H.

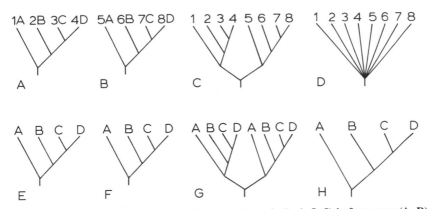

Figure 14.3. A, B. Two cladograms of four species each (1–4, 5–8) in four areas (A–D). **C, D.** Combination of the two species cladograms under the assumptions that each is a monophyletic group (C) and that each might not be a monophyletic group (D). **E, F.** Area cladograms derived from A–B. **G–H,** Possibilities for combination of area cladograms: complex in G and simple in H.

is valid as a statement of the interrelationships of the four areas. Before considering how much evidence is enough, we may first decide what kind of evidence is needed. I imagine that sufficient cladograms of this kind would be convincing. Congruence of some quantity, then, is sufficient evidence.

With these ideal cases as a background, I reformulate the basic question: Does distributional information for the world's biota supply evidence of one or more nonrandom pattern? It seems incredible that, considering what is known of the world's biota—that is, biology in general—this basic question has not already been answered in a definitive way. Again, unless I misread history, the question will be answered, because it is fundamental to our understanding of the nature and organization of the world's biota and, perhaps, to much else besides. It will be answered by people such as yourselves, who make cladograms and ponder their meaning and significance.

I have so far considered ideal cases, wherein congruence and noncongruence are easy to perceive. Real data present a number of problems, one of which is simple lack of occurrence. Although we may be interested in four areas, the group that we study may have taxa endemic only in three of them. Consider the species cladogram of figure 14.4B. This cladogram includes three species, one in each of three of the four areas of the ideal case (fig. 14.4A). The corresponding area cladogram (fig. 14.4C) is congruent with the ideal case, but it is not as informative. There are three other three-area cladograms that are congruent (fig. 14.4D–F) and eight others that are noncongruent (fig. 14.4G–P).

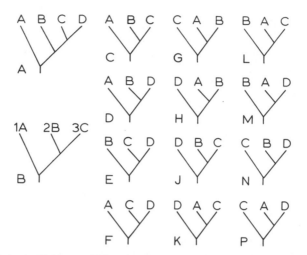

Figure 14.4. A. An ideal area cladogram wherein all areas are represented. B. Cladogram or three species (1–3) in three areas (A–C). C. Area cladogram corresponding to B. D–F. Area cladograms also congruent with A. G–P. Area cladograms noncongruent with A.

In order to appreciate the congruence or noncongruence, we may ana-lyze the cladograms according to their basic statements of relationship. Consider again species cladogram B in figure 14.4. It asserts that species 2 and 3 are related more closely to each other than either is to species 1. The corresponding area cladogram (fig. 14.4C) asserts that areas B and C are related more closely to each other than either is to area A. It is easy to see that the ideal case (fig. 14.4A) asserts the same. All we need to do is omit area D from the cladogram and observe what remains. Similarly, omit area C and achieve cladogram D; omit area A and achieve cladogram E; omit area B and achieve cladogram F. However, there is no way to achieve cladograms G–P by omission of one or more areas from cladogram A. What cladograms G–P assert is not merely different from but contrary to clado-gram A and the four less informative cladograms (fig. 14.4C–F) congruent with it.

In real data lack of occurrence is commonplace, and so are notions to explain the phenomenon. One may invoke ecology, extinction, failure to observe or to collect, or chance; in every case an educated guess is pos-sible. With respect to educated guesses, cladograms stand mute, offering no basis for preferring one guess over another. If you believe that an edu-cated guess is real knowledge, or a mark of it, you may feel slighted, as some people do, by the indifference of cladograms. If you believe that an educated guess is real biology, you may also feel that a cladogram is an empty and lifeless thing.

Because lack of occurrence reduces the information of a cladogram, we may ask where the missing information can be found. The only possibility, except to strive for ideal data for one group, is to look in a cladogram of another group. Consider two area cladograms (fig. 14.5A1, 2), from each of which one area is lacking. In combination they specify a cladogram that is fully informative for the four areas (fig. 14.5A3). Thus, lack of occurrence is a problem that can be solved by study of another group. If buttercups do not suffice, try a few liverworts. Some number of cladograms of different groups should suffice to expose the information of area relationship if it exists.

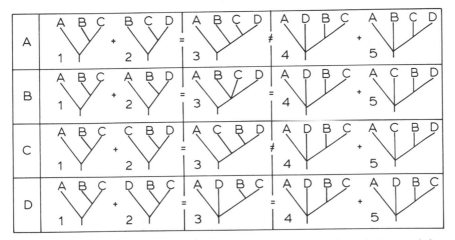

Figure 14.5. A–D. Four combinations of area cladograms. 1, 2. Area cladograms being combined. 3. Result of combination. 4, 5. Area cladograms derived under assumption 2 (to be explained later in text).

In order to understand the combination of area cladograms, we may analyze them further (fig. 14.6). Cladogram A1 says nothing about area D, which may, therefore, have its true placement anywhere in the cladogram. Basically, there are five possibilities for placement. Area D could be the "sister area" of ABC, A, BC, B, or C. Cladogram A2 says nothing about area A, which may, therefore, have its true placement anywhere in the cladogram. Again there are five possibilities. Area A could be the "sister area" of BCD, B, CD, C, or D. Thus, we have two sets of five cladograms each. Are any cladograms common to the two sets? We observe that there is one such cladogram, A3 in figure 14.5, which is the result of the combination of cladograms A1 and A2.

As I understand set theory, what we have done is to examine sets and determine their intersection. Of course, there is nothing magical about set

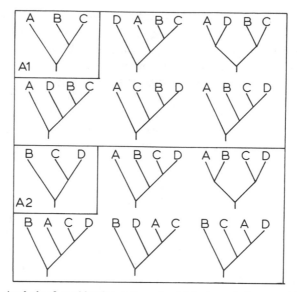

Figure 14.6. Analysis of combination A from fig. 14.5. Area cladograms A1 and A2 allow for five placements of an additional area.

theory, and set theory is not the reason why the result is correct. Cladistic theory, such as it is, dictates the result. Set theory happens to fit this aspect of cladistics.

Consider the next combination in figure 14.5 (B1–B3) and its analysis (fig. 14.7). Cladogram B1 allows for five placements of area D, and cladogram B2 allows for five placements of area C. Again we have two sets of five cladograms each. Looking for cladograms common to the two sets, we observe that there are three such cladograms, the only ones allowed by the partially informative result of the combination of cladograms B1 and B2 (fig. 14.5B3).

At this point I will introduce a notion that goes by the name of assumption 2. According to assumption 2, an area or taxon involved in a multiple basal branching may have its true placement anywhere within the structure of the cladogram. At first glance this notion seems a bit of esoterica of interest only to the crazed devotee of cladistics. In reality it is merely a way of understanding how two cladograms, incomplete in themselves, can be generally informative in combination. Because a cladogram for a real group is likely to be less than ideal—not merely because of lack of occurrence—it will always be necessary to compare cladograms that are incomplete in different ways and to extract from them what they jointly affirm, if they jointly affirm anything at all.

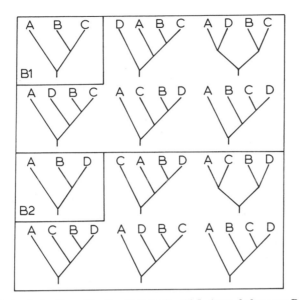

Figure 14.7. Analysis of combination B from fig. 14.5. Area cladograms B1 and B2 allow for five placements of an additional area.

Consider again cladogram A1 (see fig. 14.5) and the five possibilities for placement of area D (see fig. 14.6). Area D can be added to the basal node of the cladogram to allow for these five possibilities (fig. 14.5A4). According to assumption 2, an area or taxon involved in a multiple basal branching may have its true placement anywhere within the structure of the cladogram. In this case there are three such areas: A, D, and BC. If we view the cladogram under assumption 2, with no a priori knowledge of which area is ambiguously placed, we have to view each of the three in the same way, so that each may have its true placement anywhere. As a result the ambiguity of the multiple branching, which originally concerned only area D, is extended to areas A and BC and is, therefore, maximized. In this particular example, the gain is the inclusion of area D, which might seem small. Such, however, is not the case, for the gain is qualitative; it permits consideration of real, rather than ideal, data. It will accommodate data that are ambiguous for reasons other than lack of occurrence, which I will mention in due course. There is also a price, one that, at first glance, may seem too high. Such has been the initial impression of everyone to whom this idea has been exposed. An initial impression is not the last word, and in this case it is wrong, because the price is but a quantity, and a fairly small one at that.

If we examine the possible placements for each of the three areas, or

taxa (fig. 14.8), we find five for D, five for A, and three for BC. In sum, we
find only seven different cladograms: let us say the original five that concern D (see fig. 14.6) and the two that concern A uniquely. The price of
assumption 2, then, is the difference between five and seven. Because there
are 15 possible cladograms of this sort (dichotomous cladograms for four
taxa), the precise difference is that between $5/15$ and $7/15$—a small price to
pay for realism. The price, however, is high enough so that assumption 2
does not generally produce a fully informative result in cases of simple lack
of occurrence. I pass over the few cases for which it does give a fully informative result, for those cases are of interest only to the crazed devotee.

Consider the complementary 3-area cladogram A2 and its augmented
version, (see fig. 14.5), obtained by adding A at the basal branching under
assumption 2 (fig. 14.9). There are likewise seven different cladograms. If
we compare this set of seven with the previous set of seven, we discover
three cladograms in common, one of which is the result of our initial combination (A3) and the two others—the $2/15$.

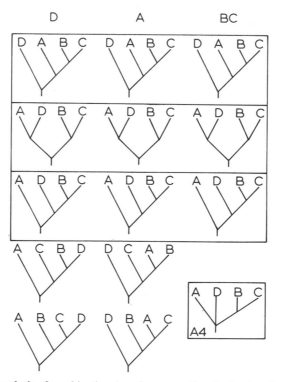

Figure 14.8. Analysis of combination A under assumption 2, showing that cladogram A4
allows for seven placements of three areas, or taxa (A, D, BC).

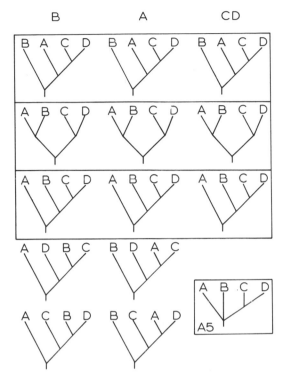

B   A   CD

Figure 14.9. Analysis of combination A under assumption 2, showing that cladogram A5 allows for seven placements of three areas, or taxa (A, B, CD).

There are numerous other possibilities of combination (see figure 14.5). In B we find that under assumption 2 the intersection of the two sets results in a partially informative result, which allows for three different cladograms; hence the result has a multiple branching that is terminal in position (fig. 14.5B3). This is as informative a result as is possible even with the a priori knowledge of what taxa are ambiguously placed (D in B4 and C in B5). In C we find a fully informative, but different, solution with a priori knowledge. In this case, as in A, the difference of ²⁄₁₅ obscures a fully informative result under assumption 2. In D we find a result that is no more informative than each of the area cladograms that are combined.

Another problem of real data is widespread taxa. Although we may be interested in four areas, the group that we study may have taxa endemic only in two areas, with a widespread taxon in the other two areas. Two such examples and their combination under assumption 2 are shown in fig. 14.10A, B. The result (fig. 14.10C) seems reasonable but arguable. Three other examples are shown in fig. 14.10D–F. The result of their combina-

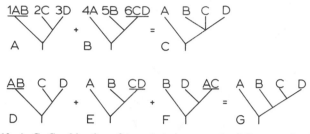

Figure 14.10. A–C. Combination of two cladograms, each of three species (1–3, 4–6) in four areas (A–D), with a partially informative result. D–G. Combination of three area clado-grams involving widespread taxa, with a fully informative result.

tion (fig. 14.10G) seems counterintuitive. I spare you the analysis of these examples (cf. Nelson and Platnick 1981), which is somewhat more com-plicated than it is for simple lack of occurrence. The point is simply that under assumption 2 sufficient cladograms, even though each is incom-plete, determine a fully informative result. I do not imply that because a result can be obtained by this or any other procedure the result is true. A test of truth is another problem.

At this point I digress in order to consider some particular widespread taxa that involve four areas: Japan, the Philippines, Queensland, and New Caledonia (figure 14.11). Shown are distributions of single species of cow-ries, genus *Cypraea*. My purpose is to illustrate that widespread taxa can be found limited to any two or more areas. The maps in figure 14.11, re-drawn from the monograph of Burgess (1970), are somewhat impression-istic but faithful to his originals. With information of this sort, we might be prone to make educated guesses about some of the variation. There are, after all, a number of possible explanations of the apparent limits of the distribution of a particular species or other taxon: taxonomic errors of lumping and splitting, failure to observe or collect, dispersal from one area to another, extinction, secretive habits, peculiar ecology, slow evolution, panmixia, homeostasis of the gene pool, etc., etc. This sort of variation, however, is commonly encountered, and an endless number of educated guesses would be needed to explain it all. In order to make progress with a cladistic approach to biogeography, one must resist the temptation to ex-plain everything by educated guesses, and even to explain any particular case in this way, for one educated guess leads to another. Again, if you feel that such explanation is real biology, you may be unwilling to give up its practice in exchange for the rather stark constraints of cladistic investiga-tion.

Widespread distributions have been of interest to biogeographers, so much

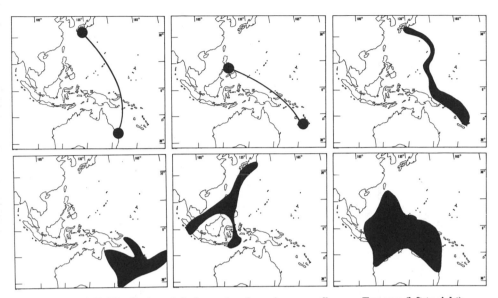

Figure 14.11. Distribution of single species of cowries, genus *Cypraea*. Top row (left to right): *C. hungerfordi*, *C. martini*, *C. guttata*. Bottom row: *C. yaloka*, *C. boivinii*, *C. quadrimaculata*. Redrawn from Burgess (1970). *C. guttata*, a rare species known from the New Hebrides and farther north, is not yet recorded from New Caledonia but is treated here as if it were.

so that one may accurately affirm that biogeographers have been preoccupied by them. In addition to explaining the distributions, biogeographers often count the different types. Just for fun I have summarized the data on the cowries in table 14.1 by the number of species found only in one area ("endemics"), as well as the number in pairs of areas (A), trios of areas (B), and all areas (C), along with the total number per area. Approximately 87 species are involved, and about 100 other species live elsewhere.

Biogeographers commonly use a mathematical index in order to assess the similarity of areas. There are various indices and various mathematical procedures for calculating similarity. If these data were analyzed according to some such procedure, perhaps the Philippines and Queensland would be found most similar, because they have six species in common, which is a number higher than that of other pairs of areas. I will natter on a little more on the subject of similarity, primarily to mention that absence as well as presence of shared taxa is sometimes used in reckonings of this sort. Thus the Philippines and Queensland are similar in having six species in common, and similar also in lacking one species that is common to the other two areas. Thus, one might reckon seven units of overall similarity between the Philippines and Queensland. For a bit of added fun I have computed

Table 14.1. Similarities (above) and overall similarities (below) of four areas, based on occurrence of species of cowries (data from Burgess 1970)

Area	Number of Endemics	Number of Widespread Species			Total
		A	B	C	
Japan	5	2 2 1 – – –	3 1 3 –	37	54
Philippines	9	2 – 6 2 –	3 – 3 3	37	65
Queensland	3	– 2 – 6 – 3	3 1 – 3	37	58
New Caledonia	7	– 1 – 2 3	– 1 3 3	37	57
Japan	8	5 4 7 – –	10 10 6 –		
Philippines	10	5 – 7 4 –	10 – 6 8		
Queensland	6	– 4 – 7 – 5	10 10 – 8		
New Caledonia	10	– 7 – 4 5	– 10 6 8		

the units of overall similarity as well (table 14.1). Here we see the revealing fact that on the whole Japan and New Caledonia are as similar, with respect to the data on pairs of areas, as are the Philippines and Queensland; in other words, $1 + 6 = 6 + 1$. With these considerations we become immersed in the conundrums of phenetics. Although a little dip into this subject might be refreshing, a prolonged submersion is bound to be debilitating. While we are in it, there is a point I wish to make by reference to an earlier figure (fig. 14.10), wherein there are area cladograms with widespread taxa. The point is that the results are independent of the number of cladograms of each type, or more precisely, of the number of widespread taxa common to two areas. As far as assumption 2 is concerned, the relative numbers of widespread taxa might just as well be random variation. The results stem only from the cladistic structure—all of it. I will not homilize on the subject of "If you feel that phenetics is real biology . . ." beyond the observation that if such is the case you have been submerged too long and you are neglecting your homework.

I grant that counting taxa common to two or more areas need not give a spurious result in any particular case. Counting genera in common, for example, allows for the possibility that the genera are natural groups of endemic species; counting species in common allows for the possibility of endemic subspecies; counting subspecies in common allows for endemic varieties; and so on, down to local demes or whatever small units might exist. At a given level, some cladistic information might be embedded in any phenetic study, and the cladistic information might determine the outcome in a particular case, providing, of course, that the information is not totally swamped by the random variables that phenetic techniques combine and weigh. Analysis of one phenetic study of 1,900 species of birds, butterflies, and bats of the Australasian area indicated close agreement between birds and butterflies, whereas the bats were anomalous (Holloway and Jardine 1968; analysis by Nelson and Platnick 1981). The discrepancy is possibly due simply to the different taxonomic practices of recognizing endemic taxa as subspecies of birds and butterflies and as species of bats. With that educated guess, which I hope will absorb what conundrums might still cling to this subject, I immediately discard it and leave it and the subject behind.

A third problem with real data is redundancy. Although we may have a group of species endemic to various areas, a given area may harbor more than one species, and the additional species may be endemic or not. The problem of redundancy thus exists with reference to both widespread and endemic taxa. In figure 14.12 three cladograms are combined, each showing the distribution of three species, two of which are widespread and partly

Figure 14.12. Combination of three area cladograms involving lack of occurrence, widespread taxa, and redundancy, with a fully informative result.

redundant. In addition, D is lacking from one cladogram and A from the other two. The result is fully informative.

If you have sympathetically followed the argument to this point, I expect that now you will suspect that something is amiss. This series of three cladograms is about as opaque to intuitive appraisal as I can devise. Still, the combination may be understood without reference to intersection of sets. If, in the first cladogram, C is related to one area more closely than to another, the only informative possibility is that C is related more closely to B than to A. If not, then the cladogram gives us no cladistic information. Similarly, if in the next cladogram C is related to some area, the possibility is B or D (one or the other but not both). Also, if in the next cladogram D is related to some area, the possibility is B or C (one or the other but not both). The last two cladograms considered together indicate that if both C and D are related to some area, the only possibility is each other. This result is the only possibility that allows each cladogram to be cladistically informative to some degree. Although we may assert correctly that the result is true in that logical sense, it need not, of course, be true in any other sense. After all, the data base of the three cladograms may in truth be either misleading or simply meaningless.

Real data are usually as opaque to intuitive appraisal as those of figure 14.12. If real data nevertheless reflect a pattern of area relationship, that pattern should be discernible. Again, I would not argue that a particular result need be true. However, if we tend to get the same result from other analyses involving the same areas, we would have some basis on which to judge the truth of the matter. The problem heretofore has been to obtain any result at all from real data.

Consider other examples of redundancy involving endemic species. First, three cladograms, each of four species, two of which occur in the same area (fig. 14.13, top). Any one cladogram can be simplified to give an informative result for three areas. I have not included the results in the figure, in the hope that they would be obvious. Next, an example, slightly more problematical, of two cladograms, each of four species, which in combination give an informative result for four areas (fig. 14.13, middle). I have omitted the result to coax you to resolve it yourself. Finally, an example, tending to be opaque to intuition, of four cladograms, each of four species,

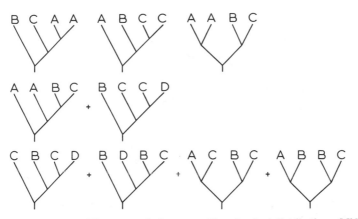

Figure 14.13. Top row: Three area cladograms with redundant distributions. Middle row: combination of two area cladograms, with fully informative result omitted. Bottom row: combination of four area cladograms, with fully informative result omitted.

which in combination give an informative result for four areas (fig. 14.13, bottom). In this case I have omitted the result out of perversity.

In summary, there are three problems with real data: lack of occurrence, widespread taxa, and redundancy. Assumption 2 can accommodate all three problems. At this time assumption 2 has not been much tested, and I am unaware of any computer procedure that puts it to work. Possibly it is defective. Maybe it is not worth testing. I present it to you, not because I wish you to test it or to adopt it, but rather to illustrate the possibilities that distribution does forever repeat and that its repetitive nature may be apparent to us in the data that we already have. I feel that these possibilities are realistic. Data may become better in the future, but the changes in data surely will be quantitative rather than qualitative improvements.

To some extent these views are at odds with some current developments in biogeography, particularly those which strive to achieve ideal data. An example was recently contributed by Edmunds (1981; see fig. 14.14). Shown are trios of genera of mayflies and mecopterans distributed in the austral temperate zone. New Zealand is always the sister area to Australia and South America. Edmunds comments on the reaction to his presentation during the 1979 vicariance biogeography symposium in New York:

> The question concerning the degree to which concordant cladograms were selected or "plucked" came up both in the auditorium and in private discussions. To allow persons to judge this in the cases cited above I present a cladogram [fig. 14.15] . . . of the members of a highly paraphyletic group considered to be one family at the time of analysis.

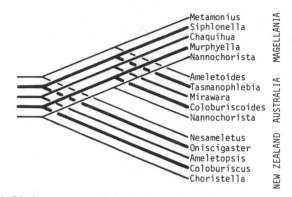

Figure 14.14. Ideal congruence of distributions of mayflies and mecopterans. After Edmunds (1981).

> Three other families—Baetidae, Caenidae, and Leptophlebiidae—remain to be analyzed. The Nannochoristidae . . . (Insecta:Mecoptera) are a *selected* example suggesting that entire cool adapted lotic water communities were vicariated. (1981:293)

Edmunds' complete cladogram (fig. 14.15) deals with four of the five groups—the mayflies, not the mecopterans—two of which show some redundancy (cf. *Rallidens, Chiloporter*), but otherwise the cladogram approaches the ideal situation. I do not suggest that ideal data are wrong, misleading, or uninteresting. Rather, citation of ideal data immediately brings to mind all of the nonideal data.

One of the few "tests" of assumption 2 was made by Platnick (1981) on data published by Rosen (1979) on two genera of freshwater fishes of Middle America (fig. 14.16). In the figure are two cladograms in what here may be considered their original form (left) and reduced form (right, following arrows). Reduction was achieved by eliminating what Rosen considered "unique components of the two area cladograms." He comments,

> The area cladograms for *Heterandria* and *Xiphophorus* differ with respect to four areas: area 3 (the Río Sarabia) which has an endemic swordtail, *S. clemenciae,* but no endemic *Heterandria* (fig. 45): and areas 6 (the Río Candelaria Yalicar), 7 (the Río Sachicha) and 9 (the Río Polochic) which have endemic species of *Heterandria* (*H. attenuata, H. cataractae,* and *H. litoperas,* respectively) but no endemic species of swordtails (figs. 46, 47. These four areas represent unique components of the two area cladograms and, as such, do not specify any general problem. . . . The area cladograms remaining after being further simplified by the deletion of unique components are congruent, i.e., they show the same areas in the same cladistic sequence (fig. 50). (p. 372)

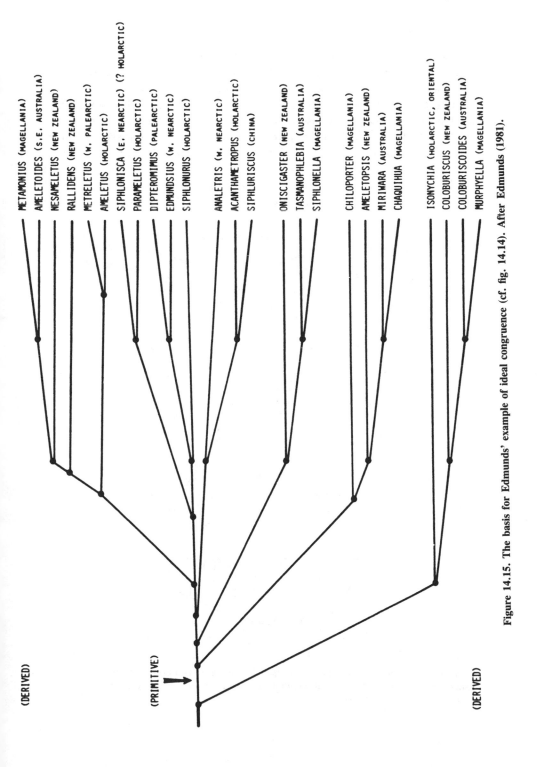

Figure 14.15. The basis for Edmunds' example of ideal congruence (cf. fig. 14.14). After Edmunds (1981).

Rosen's considerations were discussed at the 1979 vicariance biogeography symposium by Simberloff et al. (1981), who found the simplification objectionable. In effect, they argued that the simplification, or reduction, distorts a more complex, and therefore more interesting, reality, which features noncongruent sets of distributions. Indeed, the original cladograms (fig. 14.16, left) feature lack of occurrence and presence of widespread taxa; redundancy also occurs in the original data on which they are based. Under assumption 2, however, there is no noncongruence between the two cladograms, which jointly affirm a pattern of area relationship somewhat more detailed than is indicated by Rosen (fig. 14.16, below). In this case

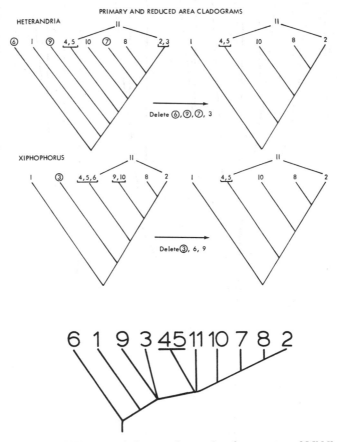

Figure 14.16. Top, middle: area cladograms for species of two genera of Middle American fishes. After Rosen (1979). Bottom: combination of the two area cladograms under assumption 2. After Platnick (1981).

assumption 2 is interesting not only because of what it reveals of cladistic structure but also because it suggests that noncongruence may be rare, perhaps nonexistent. In any case, Rosen's simplification, or reduction, in effect achieves ideal data, which anyone can appreciate as congruent.

With this I have come full circle, for here again is the problem of the general level, which begs for a decisive resolution: What about the non-ideal data, which like Lincoln's common men God must have loved, for He made so many of them. This is the real problem of biogeography, where our efforts must come together in the long run.

Literature Cited

Brundin, L. 1966. Transantarctic relationships and their significance. *Kungl. Svenska Vetenskapsakad. Handl.*, 4th ser., 11(1):1–472.

Burgess, C. M. 1970. *The Living Cowries.* Cranbury, N.J.: A. S. Barnes.

Edmunds, G. F., Jr. 1981. Discussion. In G. Nelson and D. E. Rosen, eds., *Vicariance Biogeography: A Critique.* pp. 287–297. New York: Columbia University Press.

Holloway, J.D. and N. Jardine. 1968. Two approaches to zoogeography: A study based on the distribution of butterflies, birds, and bats in the Indo-Australian area. *Proc. Linn. Soc. London* 179:153–188.

Nelson, G. and N. Platnick. 1981. *Systematics and Biogeography: Cladistics and Vicariance.* New York: Columbia University Press.

Patterson, C. 1981. Methods of paleobiogeography. In G. Nelson and D. E. Rosen, eds., *Vicariance Biogeography: A Critique*, pp. 446–489. New York: Columbia University Press.

Platnick, N. I. 1981. Widespread taxa and biogeographic congruence. In V. A. Funk and D. R. Brooks, eds., *Advances in Cladistics: Proceedings of the First Meeting of the Willi Hennig Society*, pp. 223–227. Bronx: New York Botanical Garden.

Rosen, D. E. 1979. Fishes from the uplands and intermontane basins of Guatemala: Revisionary studies and comparative geography. *Bull. Amer. Mus. Nat. Hist.* 162(5):267–376.

Simberloff, D., K. L. Heck, E. D. McCoy, and E. F. Connor. 1981. There have been no statistical tests of cladistic biogeographical hypotheses. In G. Nelson and D. E. Rosen, eds., *Vicariance Biogeography: A Critique*, pp. 40–63. New York: Columbia University Press.

Appendix

An Introduction to Computer-Assisted Cladistic Analysis

RICHARD A. ARNOLD AND THOMAS DUNCAN

One of the most important developments in systematic biology during the past 30 years has been the growth of quantitative and mathematical methods. Statistical analysis of biological data received a great impetus with the advent of electronic digital computers. Thus, it is not surprising that in recent years computers have been heavily utilized by systematists to assist in the estimation of phylogenies via cladistic methods.

Although many of the cladistic analyses undertaken by systematists are still performed quite adequately by manual techniques, large or problem data sets frequently necessitate use of a computer to derive cladograms for the groups involved. Indeed, computer-assisted analysis of cladistic data sets frequently requires less time than collection of the raw data. Execution of these analyses by computer is therefore a convenience rather than a fundamental improvement in methodology. However, the computer readily affords exploitation of cladistic techniques to their fullest extent, with very little computational tedium, and thus encourages more thorough analyses of many bodies of data. Computer-assisted cladistic techniques are particularly useful for analyzing data matrices consisting of many species and characters or those with suspected homoplasy.

Computer Structure and Function

The nature of computational hardware for scientific applications is presently undergoing a dramatic revolution in terms of size, cost, capability, and utility. During the workshop, participants' data sets were analyzed using hardware available on the Berkeley campus, including a large main-frame, a microcomputer, and two minicomputers.

Due to recent advances in storage capacity and speed of access, the modern digital computer system is capable of handling extremely complex and large-scale problems of a systematic nature. The software (i.e., programs) for large-scale computer systems is relatively easy to adapt to different machines using the same operating system. Compilers for most high-level programming languages, such as FORTRAN and PASCAL, are widely available on mini- and mainframe computers. Recently compilers for these and other programming languages have become available for microcomputers. Thus, a program written for one machine can often be adapted, with only minor modification, for use on other computer systems.

Every computer system consists of *hardware* and *software* components. Software refers to the program and data. More specifically, a program is a sequence of instructions, which directs the computer to perform particular tasks. Data are collections of alphanumeric characters manipulated by programs. Both programs and data are usually stored in files on mass storage devices. There are essentially two classes of software: systems and applications software. The former is usually provided with the computer and is required for its proper operation. It is frequently referred to as the operating system. Applications software is used to manipulate data or information.

Essential hardware components include input devices, a central storage unit, an arithmetic unit, output devices, and a control unit. Briefly, the input devices are intended to read information—data and programs—into the machine; the central storage unit, or memory, stores this information; the arithmetic unit operates on the numerical or logical data; and the output devices make the results intelligible to the user. All these processes are controlled and coordinated by the control unit.

An Introduction to Microcomputers

The term *microcomputer* is something of a misnomer, because the prefix *micro* creates the mistaken impression that a microcomputer can perform only minor tasks using small amounts of information. Despite their small physical size, for many applications microcomputers compare quite favorably with minicomputers and large mainframe computer systems. Today's microcomputers are more powerful and faster than some of the mini- and mainframe computers that were manufactured just a few years ago.

At the workshop an Apple II+ microcomputer was used by participants to analyze their respective data sets. Joseph Felsenstein kindly provided his package of cladistic programs, known as PHYLIP, in an Apple-compatible format. Approximately one-half of the participants took advantage of this situation to gain additional insight into their problems. Most were amazed at how easily they could enter their data and quickly obtain results. We hope that this experience will encourage others to use microcomputers in their research.

The microcomputer market is characterized by its variety of manufacturers, models, and options. Microcomputers are available at a reasonable cost. Excellent software for several manufacturers' systems enables scientists to perform a multitude of tasks normally associated with larger computers, e.g., data base management, sophisticated statistical analyses, simulations, and word processing. A wide spectrum of software to perform these and other tasks can be obtained through commercial and public-domain outlets. Software is becoming more "user friendly," i.e., programming experience is not necessary to utilize the computer.

The cost of a typical microcomputer system ranges from about $2,500 to $10,000, depending on the peripherals obtained. With charges for time-sharing and CPU time rising at many institutions, more systematists are turning to microcomputers as a research tool. Many research tasks, such as data analysis and management and manuscript preparation, can be accomplished less expensively with a microcomputer than with a large mainframe or time-sharing system.

Computer Programs for Cladistic Analysis

The following is a brief description of five computer programs or packages of programs for cladistic analysis. All are available at no cost or for a minimal charge. For each program we note its main function, the computing machine(s) on which it has been operated, and whom to contact regarding its availability.

1) CLINCH. Performs character compatibility analysis using the method of Estabrook. Written in FORTRAN that is compatible with a wide variety of mainframe computers. Available from Dr. Kent Fiala, Department of Ecology and Evolution, State University of New York, Stony Brook, New York 11794.

2) WAGNER-78 (and later releases). Performs Wagner parsimony analysis on both character and distance data using the method of Farris. Written in FORTRAN; can be adapted to run on a variety of mainframe computers. Details on availability may be obtained

from Dr. James S. Farris, Department of Ecology and Evolution, State University of New York, Stony Brook, New York 11794.

3) PHYLIP. A general-purpose package of programs that apply a wide variety of methods to chemical, morphological, and distance data. Methods include Fitch-Margoliash, Wagner parsimony, Camin-Sokal parsimony, character compatibility, polymorphism parsimony, Dollo parsimony, and Cavalli-Sforza and Edwards parsimony. Written in PASCAL, it is easily adapted to a variety of micro-, mini-, and mainframe computers. Distributed by Dr. Joseph Felsenstein, Department of Genetics, University of Washington, Seattle, Washington 98195.

4) EVOLVES. Performs a Fitch-Margoliash analysis from distance data. Written in FORTRAN, the code is heavily oriented toward UNIVAC mainframe computers. Can be modified for other machines with some difficulty. Available from Dr. Walter M. Fitch, Department of Physiological Chemistry, University of Wisconsin at Madison, Madison, Wisconsin 53706.

5) BIOSYS. Performs Wagner distance parsimony method on electrophoretically detectable allelic variation and computes a variety of statistics associated with genetic variability. Written in FORTRAN; easily adapted for IBM and CDC mainframe computers. Available from Dr. David L. Swofford, Department of Genetics and Development, University of Illinois at Champaign-Urbana, Urbana, Illinois 61820.

Participants in the Workshop on the Theory and Application of Cladistic Methodology

John H. Beaman
Department of Botany
Michigan State University
East Lansing, Michigan 48823

Hollie G. Bedell
Department of Botany
University of Maryland
College Park, Maryland 20742

Christopher S. Campbell
Botany Department
Rutgers University
Newark, New Jersey 07102

Nancy C. Coile
Botany Department
University of Georgia
Athens, Georgia 30602

Elizabeth A. K. Coombs
Harvard University Herbaria
22 Divinity Ave.
Cambridge, Massachusetts 02138

Kraig Derstler
Geology Department
University of California
Davis, California 95616

Michael J. Donoghue
Harvard University Herbaria
22 Divinity Ave.
Cambridge, Massachusetts 02138

Addresses of participants indicate academic affiliation at time of workshop.

Daphne F. Dunn
Department of Invertebrate Zoology
California Academy of Sciences,
San Francisco, California 94118

Wayland L. Ezell
Department of Biological Sciences
St. Cloud State University
St. Cloud, Minnesota 56301

Vicki A. Funk
New York Botanical Garden
Bronx, New York 10458

Peter Goldblatt
Missouri Botanical Garden
P.O. Box 299
St. Louis, Missouri 63166

Shirley A. Graham
Department of Biological Sciences
Kent State University
Kent, Ohio 44242

Michael C. Grant
EPO Biology
University of Colorado
Boulder, Colorado 80309

Gene Hart
Department of Botany
Washington State University
Pullman, Washington 99164

Ronald L. Hartman
Department of Botany
University of Wyoming
Laramie, Wyoming 82071

David J. Keil
Department of Biological Sciences
California Polytechnic State University
San Luis Obispo, California 93407

Alan J. Kohn
Department of Zoology
University of Washington
Seattle, Washington 98195

Sandra Lindstrom
Division of Biological Sciences
University of Michigan
Ann Arbor, Michigan 48109

Ronald J. McGinley
Museum of Comparative Zoology
Harvard University
Cambridge, Massachusetts 02138

W. Wayne Moss
Department of Entomology
Philadelphia Academy of Sciences
Philadelphia, Pennsylvania 19103

Inger Nordal
Botanical Institute
University of Oslo
Oslo, Norway

Robert D. Owen
Oklahoma Biological Survey
University of Oklahoma
Norman, Oklahoma 73019

Rhonda Riggins
Biological Sciences Department
California Polytechnic State University
San Luis Obispo, California 93407

Barry Roth
Department of Invertebrate Zoology
California Academy of Sciences
San Francisco, California 94118

Leila M. Shultz
Department of Biology
Utah State University
Logan, Utah 84322

Beryl B. Simpson
Department of Botany
University of Texas
Austin, Texas 78712

Susan M. Skillman
Department of Botany
Science Hall

Washington State University
Pullman, Washington 99164

Janet R. Sullivan
Department of Botany and Microbiology
University of Oklahoma
Norman, Oklahoma 73019

Gene S. Van Horn
Department of Biology
University of Tennessee
Chattanooga, Tennessee 37402

David B. Wake
Museum of Vertebrate Zoology
University of California
Berkeley, California 94720

Allan C. Wilson
Department of Biochemistry
University of California
Berkeley, California 94720

D. Scott Wood
Stovall Museum, Bird Division
University of Oklahoma
Norman, Oklahoma 73019

David A. Young
Department of Botany
University of Illinois
Urbana, Illinois 61801

Elizabeth Anne Zimmer
Department of Biology
Washington University
St. Louis, Missouri 63130

AUTHOR INDEX

SUBJECT INDEX

Acacia, 58, 59-61; *longifolia*, 61; *melanoxylon*, 59, 61; *verticillata*, 59, 63, 68
Acmella, 82-84
Additive Binary Coding, 126
Additivity, 222
Aizoaceae, 100
Altruism, 78
Amaranthaceae, 104
Amino acids, 222-25
Amphidiploids, 193
Anagenesis, 44
Anagenic change, 258, 268
Analogy, 53; *see also* Homology
Anastomosing branching, 193; *see also* Reticulation
Ancestor-descendant relationships, 138-40; *see also* Common ancestry
Anemia, 98; *colimensis*, 101; subg. *Coptophyllum*, 101; *villosa*, 104
Apochemical, 99
Apogamous, 99
Apomorphy, 261, 264-67; *see also* Autapomorphy, Synapomorphy
Apophenous, 99
Apoploidal, 99
Arginine, 224
Arthropoda, 39
Aspartate, 224
Aspleniaceae, 98
Asplenium, 106, 109
Australia, 289-90
Autapomorphy, 249, 261, 266; *see also* Apomorphy, Synapormorphy
Avena, 194, 196–201, 218; cultivars, 248; cultivars and species, 192; genomes (A and C), 196; *sativa*, 157, 196

Bacteria, 240-41
Balistoidea, 157

Bayes' Theorem, 177
Bear, 231
Berberidaceae, 157, 159, 160-62
Berberis, 55
Beryciformes, 157
Billia, 101
Birds, 229
Branch swapping, 240

Cactaceae, 53, 55-56, 100
Cat, 231
Catastomid fishes, 180
Centers of origin, 274
Certainty in systematics, 189
Character compatibility, 172, 181-82; analysis, 152, 183-85, 194, 196, 202, 205; calculation of probabilities, 154-56; probability analysis, 153; random, 154-56; statistical properties of, 184-85; *see also* Clique
Characters: alphanumeric, 296; basi-, 105; compatible, 195; definition of, 152; iterative weighting method, 183; logical dependence, 153; morphological, 202; probability of, 183; trends, 99; type, 260-61; weighting, 238, 250
Character-state: ambiguity, 237-38; changes, 227; correlation, 78; distribution, 260-61
Character-state trees, 139-40, 238-39; addition of true, 144-49; as hypotheses, 141-44; compatibility, 149-50
Charadriformes, 157
Cichlidae, 157
Ciliata, 43
Cladism: dichotomy, 14-15, 20; essence of, 16-17; methodological sense, 17-18; transformed, 17; transformed and untransformed, 14-15